Lecture Notes in Mathematics

Edited by A. Dold and B. Eckmann

693

Hilbert Space Operators

Proceedings, California State University Long Beach
Long Beach, California, 20–24 June, 1977

Edited by
J. M. Bachar, Jr. and D. W. Hadwin

Springer-Verlag
Berlin Heidelberg New York 1978

Editors

John M. Bachar, Jr.
Department of Mathematics
California State University Long Beach
Long Beach, CA 90840/USA

Donald W. Hadwin
Department of Mathematics
University of New Hampshire
Durham, NH 03824/USA

AMS Subject Classifications (1970): 28 A 65, 46 L 15, 47-02, 47 A 10, 47 A 15, 47 A 35, 47 B 05, 47 B 10, 47 B 20, 47 B 35, 47 B 40, 47 B 99, 47 C 05, 47 C 10, 47 C 15, 47 D 05, 47 E 05, 47 G 05

ISBN   3-540-09097-5   Springer-Verlag   Berlin   Heidelberg   New York
ISBN   0-387-09097-5   Springer-Verlag   New York   Heidelberg   Berlin

© by Springer-Verlag Berlin Heidelberg 1978
Printed in Germany

Printing and binding: Beltz Offsetdruck, Hemsbach/Bergstr.
2141/3140-543210

PREFACE

This volume contains the contributions to the Conference on Hilbert Space Operators, held at California State University Long Beach during the week of 20-24 June 1977. The purpose of the conference was to present some recent developments and some problems in Hilbert Space Operator Theory which are likely to be of importance for further advances in the field.

Three main lecturers each delivered three lectures on the main topic of concrete representations of Hilbert space operators:

1. P. R. Halmos, Integral Operators $(f(x) \to \int k(x,y)f(y)dy)$.
2. B. Abrahamse, Multiplication Operators $(f(x) \to \varphi(x)f(x))$.
3. E. Nordgren, Composition Operators $(f(x) \to f(T(x)))$.

Professor Halmos has included a description of the main topic in the introduction to his paper. Additionally, other lectures were given in the theory of Hilbert space operators, some of which are related to concrete representations of operators. The 21 papers in this volume contain, in varying degrees, historical background, expository accounts, the development and presentation of new ideas and results, and the posing of new problems for research.

The conference was funded jointly by the National Science Foundation (Grant number MCS 77-15176) and by the host, California State University Long Beach.

We express much gratitude to them for making the conference possible, to the authors for their manuscripts, to Elaine Barth for her excellent typing, to the participants, and to Springer-Verlag for publishing this volume.

John Bachar

Donald Hadwin

Conference Participants

Bruce Abrahamse
Brian Amrine
Richard Arens
Sheldon Axler
John M. Bachar, Jr.
Jose Barria
Estelle Basor
Brad Beaver
I. D. Berg
Charles A. Berger
George Biriuk
Richard Bouldin
James R. Brown
Alice Chang
Jen-Chung Chuan
Wai-Fong Chuan
Floyd Cohen
Carl Cowen
James Deddens
Charles DePrima
Les Deutsch
Henry Dye
Brent Ellerbroek
J. M. Erdman
John Ernest
John T. Fagarason
Herb Gindler
Neil Gretsky
Ted Guinn
Donald W. Hadwin
Paul Halmos
Bernard Harvey
Thomas Haskell
William Helton
Domingo A. Herrero
Michael Hoffman
Richard B. Holmes
Thomas Hoover
Donald H. Hyers
Nicholas Jewell
Darrell J. Johnson
Gerhard Kalisch
Robert Kelly
Jerry Koliha
Ray Kunze
Alan Lambert
Tan Yu Lee
Tung Po Lin
Arthur Lubin
Carl Maltz
William Margulies
John McDonald
C.R. Miers
Paul Muhly
Eric Nordgren
Catherine Olsen
Boon-Hua Ong
Joseph Oppenheim
Effrem Ostrow
Barbara Rentzsch
William G. Rosen
Mel Rosenfeld
Peter Rosenthal
James Rovnyak
Norberto Salinas
Bonnie Saunders
Howard Schwartz
Nien-Tsu Shen
Allan Shields
A. R. Sourour
Joseph Stampfli
James D. Stein, Jr.
John B. Stubblebine
Barbara Turner
Larry Wallen
Kenneth Warner
Steven Weinstein
Gary Weiss
Joel Westman
Robert Whitley
Harold Widom

CONTENTS

MAIN LECTURES

ADDITIONAL LECTURES

----

* This paper presented at the conference is not included in this volume - the results will be published elsewhere.

INTEGRAL OPERATORS

P. R. Halmos

PREFACE

The following report on integral operators, and the introduction to the three principal themes of the conference that precedes it, should be accompanied by an apology. They are not, obviously not, polished exposition; they are what at best they might seem to be, namely lecture notes. They are the notes I prepared before the conference and kept peering at as I was lecturing. In the original hand-written version there were three errors (that I know of). I corrected them, but that is the only change I made.

The result is a compressed summary for those who were not at the lectures, and a reasonably representative reminder for those who were. I have agreed to its publication now, so that the proceedings of the conference may have some claim to completeness. A more detailed exposition of the part of the theory of integral operators that I am interested in will be contained in a research monograph that is now in preparation.

INTRODUCTION -

The major obstacle to progress in operator theory is the dearth of concrete examples whose properties can be explicitly determined.

All known (and perhaps all conceivable) examples belong to one of three species. The reason for that is that the only concrete example of Hilbert space is L^2 (over some measure space), and there isn't much one can do to the functions in L^2.

The simplest thing to do is to fix a function φ and multiply every f in L^2 by that φ. (In order for this operation to turn out to be bounded, the multiplier φ must, of course, belong to L^∞.) The multiplication operators so obtained, and their immediate family, are the best known and most extensively studied examples. The spectral theorem assures us that every normal operator is of this kind. The special case of diagonal matrices is too easy to teach us much but is, nevertheless, too important to be neglected. Multiplicity theory, unitary equivalence theory, and the effective calculability of invariant subspaces of diagonal matrices can more or less be extended to all multiplication operators, and a large part of operator theory is directed toward making it more instead of less.

Dilation theory began with the observation that (to within unitary equivalence) every operator can be obtained by compressing multiplication operators to suitable subspaces. Certain special compressions (for example, the restrictions to invariant subspaces, which yield the subnormal operators, and, for another example, the ones suggested by the passage from certain groups to their most important subsemigroups, which yield the Toeplitz operators) are amenable to study. M. B. Abrahamse has made substantial contributions to several aspects of the theory of multiplication operators, broadly interpreted, and in his lectures will present a part of that theory.

Next to multiplication the simplest thing to do to a function is substitution: to get a new function from an old one, calculate the value of the old function at a new place. In symbols: map $f(x)$ to $f(Tx)$. Simplest instance: map a sequence $\{f(n)\}$ in ℓ^2 to the translated sequence $\{f(n+1)\}$, thus getting a shift (unilateral or bilateral). More sophisticated: let T be a measure-preserving transformation on the underlying measure space, and thus make contact with ergodic theory.

Members of the same species can interbreed; combinations of multiplication operators and substitution operators yield weighted shifts and, more generally, weighted translation operators, studied by Parrott and others.

If the underlying space has additional structure (e.g., analytic structure), and the substitutions permitted are correspondingly richer, the theory makes contact with classical analysis. Much of this circle of ideas has been studied by E. A. Nordgren.

In a sense the most natural, but, as it turns out, the least helpful way to try to construct operators is via infinite matrices — with rare exceptions (diagonal, Toeplitz, Hankel), which can usually be subsumed under multiplications and substitutions, matrices have not been a rich source of examples. Integral kernels are generalizations of matrices, and, incidentally, are the source of almost all modern analysis. I turned to them a few years ago in the hope of finding rewarding examples, and found that they have quite an extensive theory that is not yet completely worked out — there are still reasons to maintain the hope of reward. I shall try to tell you something about the present state of the theory of integral operators.

LECTURE 1. CONCEPTS

Definitions.

An integral operator is induced by a measurable function k on the Cartesian product $X \times Y$ of two measure spaces by an equation such as

$$f(x) = \int k(x,y)g(y)dy .$$

Known isomorphism theorems in measure theory make it possible to ignore all but the pleasantest of classical measure spaces with no essential loss of generality: the only spaces that need to be considered are the finite $\mathbb{Z}_n$ $(=\{1,\ldots,n\})$ and $\mathbb{I}$ $(=[0,1])$, and the infinite $\mathbb{Z}$ (or $\mathbb{Z}_+$) and $\mathbb{R}$ (or $\mathbb{R}_+$). I shall refer to $\mathbb{Z}_n$ and $\mathbb{Z}$ as the atomic cases and to $\mathbb{I}$ and $\mathbb{R}$ as the divisible cases.

In addition to the isomorphism theorems that leave measures as they find them, there are some important ones that change measures. Their use makes it possible to pass back and forth between finite and infinite spaces; the only really important distinction is the one between atomic and divisible spaces. The theory I am hinting at is perfectly illustrated by the effect of the mapping $\varphi : \mathbb{I} \to \mathbb{R}_+$,

$$\varphi(x) = \frac{x}{1-x} .$$

The transformation $U : L^2(\mathbb{R}_+) \to L^2(\mathbb{I})$ defined by

$$Uf(x) = \frac{1}{\sqrt{\delta(\varphi(x))}} f(\varphi(x)),$$

where δ is the (Radon-Nikodym) derivative, $\delta(x) = \dfrac{1}{(1+x)^2}$, is unitary, it sends integral operators to integral operators, and it preserves all properties of kernels pertinent to the category (operators on Hilbert space) under study.

At the heart of the theory is the finite divisible case, i.e., the unit interval. In the systematic search for examples, however, it is unwise to ignore the atomic case ($\mathbb{Z}$) and the infinite case ($\mathbb{R}$). There is a standard "inflation" process for converting a matrix a (on $\mathbb{Z}_+ \times \mathbb{Z}_+$) into a kernel k (on $\mathbb{R}_+ \times \mathbb{R}_+$): write $k = a(i,j)$ on the unit square with diagonal from (i,j) to $(i+1, j+1)$. The change of measure described before can be used to get examples over $\mathbb{I}$ from examples over $\mathbb{R}_+$; combined with matrix inflation, it can be used to get examples over $\mathbb{I}$ from examples over $\mathbb{Z}_+$. In what follows I shall be interested in $\mathbb{I}$ only, and when I say "kernel" I shall mean "kernel on $\mathbb{I} \times \mathbb{I}$"; in view of the preceding comments, however, when I need an example of a kernel, I shall feel free to produce one elsewhere and expect that inflation and change of measure will be applied automatically so as to re-establish contact with $\mathbb{I}$.

An <u>integral operator</u> is one induced by a kernel; in order for that expression to make sense the kernel has to satisfy the following three conditions:

(1) If $g \in L^2$, then $k(x,\cdot)g \in L^1$ almost everywhere.

(2) If $g \in L^2$, then $\int k(\cdot,y)g(y)dy \in L^2$.

(3) There exists a constant c such that if $g \in L^2$, then $\| \int k(\cdot,y)g(y)dy \| \leq c\|g\|$.

Banach knew 45 years ago that (1) and (2) imply (3); this useful fact follows from some non-trivial measure theory (and not just from straightforward application of the closed graph theorem).

What are some typical examples of integral operators? The best known ones are the Hilbert-Schmidt operators induced by kernels in L^2. They are compact. Among the simplest kernels in L^2 are the ones of the form

$$k(x,y) = u(x)v(y)$$

(with u and v in L^2); the corresponding operators have rank 1. One of the simplest of <u>them</u> is given by

$$k(x,y) \equiv 1;$$

the corresponding operator is a projection of rank 1.

Compactness counterexamples.

Because Hilbert-Schmidt operators continue to play an important part in the theory, there is some popular confusion about the role of compactness in the study of integral operators. To eliminate any possibility of confusion here, I proceed to give two examples: first an integral operator that is not compact, and second a compact operator on L^2 that is not integral.

The identity operator on an infinite-dimensional Hilbert space is not compact, and hence the infinite identity matrix is a kernel (atomic case) whose induced operator is not compact. The inflated kernel on $\mathbb{R}_+ \times \mathbb{R}_+$ no longer induces the identity operator; it induces a projection of infinite rank and infinite co-rank, which is, therefore, not compact either. Change measure so as to squeeze $\mathbb{R}_+$ into $\mathbb{I}$ and obtain a kernel k on $\mathbb{I} \times \mathbb{I}$ that induces a projection of infinite rank and infinite co-rank. (The kernel k is given by

$$k(x,y) = \frac{1}{(1-x)(1-y)}$$

when (x,y) is in one of the squares with diagonal from $(0,0)$ to $(1/2, 1/2)$, $(1/2, 1/2)$ to $(2/3, 2/3)$, $(2/3, 2/3)$ to $(3/4, 3/4)$, etc., and is 0 elsewhere. The functions that are $\frac{1}{1-x}$ between $\frac{n}{n+1}$ and $\frac{n+1}{n+2}$ and 0 elsewhere are an orthogonal but not normalized basis for the range of the induced projection.) So much for that: integral operators need not be compact.

The other kind of example is much harder to establish: what techniques can there possibly be for proving that something is <u>not</u> an integral operator? It's not that the example is complicated, but the proof that it works is not on the surface.

If $\{e_n\}$ and $\{e'_n\}$ are orthonormal bases in L^2, and if $\{\lambda_n\}$ is a sequence of numbers with $\lambda_n \to 0$, then there is a unique operator C such that $Ce_n = \lambda_n e'_n$; the operator C is the prototypical compact operator. If $\{e_n\}$ is pointwise uniformly bounded (e.g., the usual exponentials, re-arranged so as to form a unilateral sequence), if $\{e'_n\}$ is arbitrarily large almost everywhere, and if the λ_n's are not too small, then C is not an integral operator. (To say that

$\{e_n'\}$ is arbitrarily large almost everywhere means that $\sup_n |e_n'(x)| = \infty$ almost everywhere.) A suitable choice for the λ_n's is $\lambda_n = 1/\sqrt{\|e_n'\|_\infty}$. An example of the latter kind of basis is given by the Haar functions. They come in batches of 2^n, each one has a single dyadic interval for support, and is positive and negative (in fact $\pm\sqrt{2^n}$) on half that interval.

A necessary condition on integral operators can be obtained as follows. If k is a bounded kernel, then

$$\Omega(x) = \int |k(x,y)|\,dy < \infty$$

almost everywhere (because $1 \in L^2$). The function Ω is measurable, finite, positive, and is such that if $g \in L^\infty$, then

$$|(\text{Int } k)g(x)| \leq \|g\|_\infty \cdot \Omega(x)$$

almost everywhere. This observation is due to my student V. S. Sunder; it is inspired by a similar technique used for a similar purpose at another part of the theory by V. B. Korotkov. These two names should be referred to frequently in what follows, but I intend to let this one mention do for all; much of the theory I shall describe was discovered (within the last ten years, and some within the last six months) by one or another of these two mathematicians (and sometimes by both).

For the operator C no such Ω can exist. Reason: $|Ce_n(x)| = \lambda_n |e_n'(x)|$ is arbitrarily large almost everywhere.

Questions.

There are two basic problems in integral operator theory (which operators are integral, and which operators can be integral?), and the preceding discussion touches one of them. By "can be", I mean "are unitarily equivalent to". The Sunder-Korotkov condition is necessary for an operator to be integral. No usable necessary and sufficient condition is known; the characterization problem for integral operators is unsolved. There may be a sense in which it cannot be solved in general; good answers are known for some special classes of operators. Note, however, that the condition, while "natural" in the category of L^2 spaces (it refers to the underlying measure structure) is not natural in the category of Hilbert spaces.

It is conceivable that some operator unitarily equivalent to the counterexample C above is an integral operator, even though C itself is not. It is not only conceivable: it is true. It will turn out later, as a special case of the general theory of integral operators, that every compact operator <u>can be</u> integral.

Consider, by way of contrast, the projection of infinite rank and infinite co-rank that was proved to be an integral operator above, and compare it with the multiplication operator induced on L^2 by the characteristic function of the first half of the unit interval. Since the latter, too, is a projection of infinite rank and infinite co-rank, the two are unitarily equivalent. Is the multiplication projection an integral operator? The answer will turn out to be no.

What about more general multiplication operators? Are some of them integral operators? Which ones? And what about the ones that are not: can they be (unitarily equivalent to) integral operators? What about concrete special cases such as the position operator on $\mathbb{I}$? What about well-behaved special cases, e.g., the invertible ones (induced by multipliers bounded away from both 0 and ∞)?

For a final teaser: what about shifts? Is the unilateral shift (of multiplicity 1) unitarily equivalent to an integral operator? What about other multiplicities, up to and including aleph null?

LECTURE 2. ALGEBRA

<u>Absolutely bounded and Carleman kernels</u>.

The flavor of the theory of integral operators is greatly affected (both historically and conceptually) by two important special classes of kernels: the absolutely bounded ones and the Carleman kernels.

A kernel k is <u>absolutely bounded</u> ($k \in \mathfrak{a}$) if $|k|$ is bounded. A <u>Carleman kernel</u> is one for which $k(x,\cdot) \in L^2$ for almost every x.

The definition of boundedness for kernels involves (absolute) Lebesgue integrability: it may be surprising to learn that some kernels are not absolutely bounded. Here is an example: $X = \mathbb{Z}$, $Y = \mathbb{I}$, $k(n,y) = e^{-2\pi iny}$. This can be replaced by a divisible example (on $\mathbb{R} \times \mathbb{I}$ instead of $\mathbb{Z} \times \mathbb{I}$) by inflation, and squeezed into a finite measure space (on $\mathbb{I} \times \mathbb{I}$) by change of measure.

A classical example is given by the Hilbert matrix $(a(i,j) = \frac{1}{i-j}$ if $i \neq j)$. A quick way to prove that it is bounded but not absolutely bounded is to use Toeplitz theory. Inflation and change of measure can convert this to an example on $\mathrm{I\!I} \times \mathrm{I\!I}$.

Another example, one that is adaptable to several purposes, is obtained as a direct sum of finite matrices. For each positive integer p, let ω_p be a primitive p-th root of 1, and let w_p be the $p \times p$ matrix defined by $w_p(m,n) = \omega_p^{mn}$. A direct computation shows that $\frac{1}{\sqrt{p}}\, w_p$ is unitary and $\frac{1}{p}\,|w_p|$ is a projection. It follows that if w is the direct sum of all $\frac{1}{\sqrt{p}}\, w_p$'s, then w is unitary, but $|w|$ is not bounded. A slight modification of the argument shows that the direct sum of all $\frac{1}{p^{3/4}}\, w_p$'s is compact and still not absolutely bounded. Inflation and change of measure converts these examples to ones with the same properties on $\mathrm{I\!I} \times \mathrm{I\!I}$.

The unitary example w above not only fails to be absolutely bounded, but, in fact, is far away from every absolutely bounded kernel: if a is absolutely bounded, then $\|w - a\| \geq 1$. Consequence: the set $\mathcal{A}$ is nowhere dense in the set of all bounded kernels.

The set $\mathcal{A}$ is small enough to avoid the pathology of the subject and yet large enough for much useful algebra and analysis. Typical results: if k is absolutely bounded and if $k^*(y,x) = \overline{k(x,y)}$, then k^* is absolutely bounded and $\operatorname{Int} k^* = (\operatorname{Int} k)^*$; if k_1 and k_2 are absolutely bounded, then $k_1(x,\cdot)k_2(\cdot,y) \in L^1$ for almost every $\langle x,y \rangle$, and if, moreover,

$$k(x,y) = \int k_1(x,z)k_2(z,y)dz ,$$

then k is absolutely bounded and $\operatorname{Int} k = (\operatorname{Int} k_1)(\operatorname{Int} k_2)$.

It is a well-known pretty fact in analysis that the only functions that multiply L^2 into L^1 are the ones in L^2. The definition of a bounded kernel seems to say that $k(x,\cdot)$ multiplies L^2 into L^1 for almost every x, and hence that k is a Carleman kernel. Appearances are deceptive and the conclusion is wrong. The difficulty is one of too many sets of measure zero not adding to a set of measure zero. In fact there exist kernels that are not only bounded but, in fact, absolutely bounded but that are, nevertheless, not Carleman kernels. A classical example was used by Abel in his study of the vibrating string:

$$k(x,y) = \begin{cases} \dfrac{1}{\sqrt{x-y}} & \text{if } x > y, \\ 0 & \text{otherwise}. \end{cases}$$

The Fourier example $(e^{-2\pi i n y})$ shows that a Carleman kernel may fail to be absolutely bounded.

Adjoints and products.

The Fourier example shows something else that can go seriously wrong. If $k(n,y) = e^{-2\pi i n y}$ and $A = \text{Int } k$, then A^* assigns to each sequence g in ℓ^2 the function whose sequence of Fourier coefficients it is. Is it true that $A^* = \text{Int } k^*$? No, it's as false as can be: k^* is not a bounded kernel, and A^* is not an integral operator. (If $\Sigma_n k(x,n)g(n) = (\Sigma_n g(n)e_n)(x)$ whenever $g \in \ell^2$, then let g be the characteristic function of $\{m\}$ and conclude that $k(x,m) = e^{2\pi i m x}$. If this k were bounded, then every ℓ^2 sequence would belong to ℓ^1.)

The bad news is partially compensated for by an elegant theorem: the two things that can go wrong necessarily happen together. That is: if the adjoint of an integral operator is an integral operator at all, then the kernel that induces the adjoint must be the conjugate transpose of the original kernel, and if the conjugate transpose of a bounded kernel is also bounded, then the operator it induces must be the adjoint of the one that the original kernel induces. For Carleman kernels this theorem has been known for some time; in the general case it was observed only recently (by Sunder).

The idea of the proof (for both parts of the theorem) is to approximate the kernel by "smaller" kernels for which the desired conclusion is easier to prove. To illustrate the technique, I'll sketch the proof of the assertion that if both $A = \text{Int } k$ and $B = \text{Int } k^*$ are bounded, then $B = A^*$.

Since $1 \in L^2$, therefore $k(x,\cdot) \in L^1$ for almost every x, and hence

$\Omega(x) = \int |k(x,y)|dy < \infty$ for almost every x. If $F(N) = \{x : \Omega(x) \leq N\}$, then $\{F(N)\}$ is a sequence of sets increasing to X as N tends to ∞. The set of f's in $L^\infty(X)$ for which support $f \subset F(N)$ (for some N) is dense in $L^2(X)$, and the set of g's in $L^\infty(Y)$ is dense in $L^2(Y)$; for f's and g's of that kind the function h defined by

$$h(x,y) = f(x)k(x,y)g(y)$$

is in $L^1(X \times Y)$. For f's and g's of that kind, moreover,

$$\begin{aligned}(A^*f,g) = (f,Ag) &= \int f(x)\left(\int \overline{k(x,y)g(y)dy}\right)dx \\ &= \int\left(\int \overline{k(x,y)}f(x)dx\right)g(y)dy \\ &= (Bf,g)\,.\end{aligned}$$

Since the extreme terms in this chain of equations depend continuously on f and g, it follows, as promised, that $B = A^*$.

So much for adjoints; what about products? Nothing is known about the general case, but for Carleman kernels all goes well. The assertion is that the set of all Carleman operators is not only closed under multiplication, but, in fact, is a right ideal in the algebra of all operators. Here is a dishonest proof. Suppose that A is a Carleman operator, so that

$$Ag(x) = (g,\overline{k(x,\cdot)})\,;$$

then

$$ABg(x) = (Bg,\overline{k(x,\cdot)}) = (g,B^*\overline{k(x,\cdot)})\,;$$

it follows that AB has a kernel h, namely the one defined by

$$h(x,y) = \overline{B^*\overline{k(x,\cdot)}(y)}\,.$$

The reason this is dishonest is that $B^*\overline{k(x,\cdot)}$ is determined only to within a set of measure zero, which may vary with x. There are measurability difficulties, which must be, and can be, circumvented, but the germ of the proof remains the same.

The identity operator.

The best known operator that is not an integral operator is the identity (on $L^2(\Pi)$). The proof is not difficult but neither is it obvious: it requires either a trick or some technique. Two partial results are illuminating.

One: The identity is not an integral operator with a square-integrable kernel. Reason: such a kernel yields a Hilbert-Schmidt operator, and hence a compact one, and the identity is not compact.

Two: the identity is not an integral with a "semi-square-integrable" kernel — an ad hoc expression intended to mean "Carleman kernel". The reason is an elegant theorem of Korotkov: an operation A on L^2 is induced by a Carleman kernel if and only if there exists a measurable, finite, positive function Ω such that

$$|Ag(x)| \leq \|g\|_2 \cdot \Omega(x)$$

almost everywhere. Only if: $Ag(x) = \int k(x,y)g(y)dy$, with $k(x,\cdot) \in L^2$ almost everywhere, implies that $|Ag(x)| \leq \sqrt{\int |k(x,y)|^2 dy}\sqrt{\int |g(y)|^2 dy}$. The proof of "if" is harder, but a dishonest sketch (which can be made honest) looks like this. The assumption says [?] that, for fixed x, the mapping $g \mapsto Ag(x)$ is a bounded linear functional, and hence that $Ag(x) = (g,k_x)$ for some k_x in L^2; define $k(x,y)$ so that $k_x(y) = k(x,y)$. [Trouble: $Ag(x)$ doesn't mean very much at any individual point x, and "choosing" $k(x,\cdot)$ as one of the determinations of k_x leads to grave measurability difficulties. Both troubles can be averted by using the separability of L^2 and applying the sketched argument to the terms of an orthonormal basis only.]

Once Korotkov's theorem is available, the assertion that the identity is not a Carleman operator becomes transparent: no finite-valued function can pointwise dominate (almost everywhere) all the functions in the unit ball of L^2.

There is another question in this circle of ideas that I have not emphasized but that deserves mention: which operators on L^2 must be Carleman operators in the sense that everything unitarily equivalent to one of them is a Carleman operator? Example: any Hilbert-Schmidt operator. Theorem: that's all — a necessary and sufficient condition that an operator be a "universal" Carleman operator is that it be a Hilbert-Schmidt operator.

LECTURE 3. ANALYSIS

Von Neumann's techniques.

The hardest and most interesting question is the one that asks which operators can be integral operators. The first systematic attack was made by von Neumann in 1935 for Hermitian Carleman operators. (The extension by Weidmann in 1970 to all Carleman operators was relatively easy.)

Von Neumann's main contributions to this field are a beautiful theorem and a simple but beautiful trick.

The theorem is that Hermitian operators on infinite-dimensional spaces come very near to being as diagonalizable as their finite-dimensional cousins: if A is Hermitian, then $A = D + S$, where D is Hermitian and diagonalizable and S is a Hilbert-Schmidt operator. A relatively recent and relatively short and transparent proof of von Neumann's theorem uses the theory of quasidiagonal operators. (An operator is quasidiagonal if it is $B + C$, where B is block-diagonalizable and C is compact. Useful technical theorem: the set of quasi-diagonal operators is closed. Consequence via the spectral theorem: every Hermitian operator is quasidiagonal. The von Neumann theorem follows from an only moderately delicate matrix argument.)

The trick is based on the observation that an infinite-dimensional 0 is the direct sum of infinitely many copies of itself. It follows that if A is Hermitian, there is no loss of generality in writing $A \oplus 0 = (B \oplus 0 \oplus 0 \oplus \cdots) + (S \oplus 0 \oplus 0 \oplus \cdots)$, where B is Hermitian and diagonal [!] and S is a Hilbert-Schmidt operator. Rewrite $B \oplus 0 \oplus 0 \oplus \cdots$ as $B_1 \oplus B_2 \oplus \cdots$ where each B_i has rank 1 (adjoin each diagonal element to a different 0). Since each B_j is a Hilbert-Schmidt operator, it is a (Carleman) integral operator, and therefore the direct sum of the B_j's is a Carleman operator. The summand $S \oplus 0 \oplus 0 \oplus \cdots$ is trivially a Carleman operator, and, therefore, so is the sum $A \oplus 0$.

What distinguishes the Hermitian operators of the form $A \oplus 0$ among all Hermitian operators? From the present point of view the answer is that the number 0 not only belongs to the spectrum of every direct sum such as $A \oplus 0$ but (because of the infinite-dimensionality of the 0 direct summand) continues to belong even if $A \oplus 0$ is subjected to an arbitrary compact perturbation. Equivalently: not only does the equation $(A \oplus 0)X = 1$ fail to have any solution, but the same is true of the equation $(A \oplus 0)X = 1 + C$, where C is compact. In technical language: 0 belongs to the essential spectrum of $A \oplus 0$. A perturbation of the proof given above establishes von Neumann's conclusion: if A is Hermitian and has 0 in its essential spectrum, then A is a Carleman operator.

Right essential spectrum.

What about operators that are not Hermitian? By the right-ideal theorem, if A is a Carleman operator, then so is AX for all X; which operators have the form AX with A Hermitian and 0 in the essential spectrum? Answer (via polar decomposition): all operators for which 0 belongs to the right essential spectrum.

How near is the sufficient condition so obtained to a necessary one? Could it be that every integral operator on $L^2(\mathrm{II})$ has 0 in its right essential spectrum? The answer is yes; that is the main result of Sunder and Korotkov. (For the special case of Hermitian Carleman operators the answer was discovered by von Neumann.)

The proof is not too bad but it does take quite a bit of analysis; I content myself here with mentioning the two main tools: a uniform absolute continuity theorem and a compactness theorem.

For each set E of positive measure in II the multiplication operator induced by the characteristic function of E is a projection P_E. For each g in L^2, the norm $\|P_E g\|$ is small when $\mu(E)$ is small, but not uniformly so: the way $\|P_E g\|$ varies with E depends very much on g. Indeed if $E \mapsto \|P_E g\|$ were "uniformly absolutely continuous", then $E \mapsto \|P_E\|$ would be "absolutely continuous", i.e., $\|P_E\|$ would become small when $\mu(E)$ did, which is patently absurd — in fact $\|P_E\|$ is identically equal to 1.

Here is where an important new aspect of the theory enters: since $L^2 \subset L^1$, each operator A on L^2 induces an operator $A_{2,1}$ from L^2 into L^1. The preceding paragraph showed that if $A = 1$, then A is <u>not</u> "absolutely continuous" (in the sense that $\|AP_E\|$ is not small when $\mu(E)$ is). Assertion: $A_{2,1}$ is "absolutely continuous". Verification: if $g \in L^2$, then $\|A_{2,1} P_E g\|_1 = \int |\chi_E g| d\mu = (\chi_E, |g|) \leq \|\chi_E\| \cdot \|g\| = \sqrt{\mu(E)} \ \|g\|$, so that $\|A_{2,1}\| \leq \sqrt{\mu(E)}$.

This kind of "absolute continuity" does not seem to be fully understood as yet, but what is known and useful is that it is true for every integral operator. That is the uniform absolute continuity theorem.

Compactness theorem.

The compactness theorem might come as a bit of a surprise. I emphasized before that compact operators can fail to be integral, and, more important, integral operators can fail to be compact. Despite that emphasis, I now report that in a certain sense all integral operators must be compact; the sense is that if A is an integral operator, then $A_{2,1}$ is compact.

For which operators is the last assertion true? How large is the set $\mathcal{K}_{2,1}$ of all operators A on L^2 for which $A_{2,1}$ is compact? If C is compact, then C is in $\mathcal{K}_{2,1}$: that's obvious. It is also easy to see that the set $\mathcal{K}_{2,1}$ is closed under the vector operations, and closed in the norm topology. If, moreover, A is in $\mathcal{K}_{2,1}$ and B is an arbitrary operator on L^2, then $(AB)_{2,1} = A_{2,1} \cdot B$, and therefore AB is in $\mathcal{K}_{2,1}$. Conclusion: $\mathcal{K}_{2,1}$ is a closed right ideal.

Assertion: the identity operator is not in $\mathcal{K}_{2,1}$. Reason: if $\{e_n\}$ is the usual exponential basis (almost any other orthonormal basis would do here), then $e_n \to 0$ weakly in L^2, but it is not true that $1 \cdot e_n \to 0$ strongly in L^1. (A small modification of the proof shows that if A is a non-zero multiplication operator on L^2, then A is not in $\mathcal{K}_{2,1}$.)

The characterization of integral operators is now within reach. If A is an integral operator, so that A is in $\mathcal{K}_{2,1}$, then $AB - C$ is in $\mathcal{K}_{2,1}$ whenever B is an arbitrary operator and C is compact. Since 1 is not in $\mathcal{K}_{2,1}$, it follows that $AB - 1$ cannot be compact: in other words 0 is in the right essential spectrum of A.

With the techniques now available many of the questions raised above can be answered. Thus, for example: although some compact operators on L^2 are not integral, nevertheless, since they all have 0 in the essential spectrum, they all can be. An invertible operator (which does not have 0 in its spectrum, essential or not) cannot be integral. The unilateral shift U has a right inverse (namely U^*) modulo compact operators, hence it cannot be integral, and the same is true for shifts of all finite multiplicities; the shift of infinite multiplicity can be integral.

- 0 -

Ite, missa est.

RESEARCH PROBLEMS?

(1) Is every (possibly non-measurably induced) integral operator induced by a measurable kernel?

(2) Does an integral operator "effectively" determine its kernel?

(3) Is every kernel with closed domain closed?

(4) Does every operator have an absolutely bounded matrix?

(5) Is the tensor product of two bounded kernels bounded?

(6) If the adjoint of a Carleman operator is an integral operator, is it a Carleman operator?

REFERENCES

V. B. Korotkov, Strong integral operators, Math. Notes 16 (1974), 1137-1140.

V. S. Sunder, Characterization theorems for integral operators, Indiana University Dissertation (1977).

J. von Neumann, Charakterisierung des Spektrums eines Integraloperators, Hermann, Paris (1935).

J. Weidmann, Carlemanoperatoren, Manuscripta Math. (1970), 1-38.

14 July 1977

SANTA BARBARA, CA

MULTIPLICATION OPERATORS

M. B. Abrahamse*

This expository paper considers the unitary equivalence problem within the class of multiplication operators. My intention is to present a solution of this problem due to T. L. Kriete and myself [1] which is accessible to beginners and to workers in other fields. The development makes use of direct integrals and the theorem on disintegration of measures, two tools used in a variety of areas in analysis. These tools are motivated by examples and stated precisely, although proofs are referred to in the literature. The paper also draws upon the folklore of operator theory. In this regard, I would like to mention in particular my teacher, Ronald Douglas, and my colleague, Tom Kriete; the ideas I have learned from these two people are sprinkled liberally throughout.

The paper is divided into the following seven sections.

1. Examples.
2. Direct integrals.
3. Disintegration of measures.
4. The direct integral for a multiplication operator.
5. The essential pre-image.
6. Examples revisited.
7. Observations.

The problem is stated precisely in Section 1 and the nature of the problem is explored by means of several examples. Certain theorems are presented in Sections 2 and 3 which are used to give a general solution to the problem in Sections 4 and 5. The general solution is applied to the original set of examples in Section 6 and some closing comments are made in Section 7.

1. EXAMPLES.

An operator A on a Hilbert space $\mathcal{H}$ is said to be unitarily equivalent to

*While preparing this paper, the author was supported in part by National Science Foundation Grant GP-MPS-75-04594.

an operator B on a Hilbert space $\mathcal{K}$, denoted $A \simeq B$, if there is a unitary operator U from $\mathcal{H}$ onto $\mathcal{K}$ such that $UA = BU$. If A and B are unitarily equivalent, then anything that can be said in the language of Hilbert space about A can also be said of B and conversely. Let X be a locally compact separable metric space, let μ be a sigma-finite Borel measure on X, let φ be in $L^\infty(\mu)$, and let M_φ be the multiplication operator on $L^2(\mu)$ defined by the equation $M_\varphi(f) = \varphi f$. Let Y be a second locally compact metric space, let ν be a sigma-finite Borel measure on Y, and let ψ be in $L^\infty(\nu)$. This paper considers the following problem: When is M_φ unitarily equivalent to M_ψ? A general solution to this problem is presented in Sections 2 through 5. In this section we consider six examples which suggest the general result.

Before looking at these examples, let us observe one elementary fact about the general problem. If M_φ is unitarily equivalent to M_ψ, then the essential range of φ is equal to the essential range of ψ. This is because the essential range of φ can be described in Hilbert space terms as the set of complex numbers λ such that $M_\varphi - \lambda$ is not invertible (the spectrum of M_φ). The reader should have no difficulty proving this; if he does, he can consult the hints or the solution in [6, Problem 52]. Because of this fact, in each example below, the essential range of φ is equal to the essential range of ψ.

Example 1.

Let $X = [0,1]$, $d\nu(x) = dx$, $\varphi(x) = x$, $Y = [0,1]$, $d\nu(x) = x^2 dx$, and $\psi(x) = x$. Define $U : L^2(\mu) \to L^2(\nu)$ by the equation $U(f)\,(x) = \frac{1}{x}\, f(x)$. It is readily verified that U is unitary and $UM_\varphi = M_\psi U$, hence, $M_\varphi \simeq M_\psi$.

Example 2.

Let $X = [0,1]$, $d\mu(x) = dx$, $\varphi(x) = x$, $Y = [0,1]$, and $\psi(x) = x$. Let $\{r_n : n = 1,2,\ldots\}$ be an enumeration of the rational numbers in $[0,1]$ and set

$$\nu(E) = \sum_{r_k \in E} 2^{-k}. \tag{1.1}$$

Suppose that U is an operator satisfying $M_\psi U = UM_\varphi$. Then $M_\psi^2 U = M_\psi UM_\varphi = UM_\varphi^2$ and an elementary induction argument gives $M_\psi^k U = UM_\varphi^k$ for any positive integer k.

It follows that for any polynomial p,

$$p(M_\psi)U = Up(M_\varphi)\,. \tag{1.2}$$

Fix a positive integer k and a function f in $L^2(\mu)$. Equations (1.1) and (1.2) imply that for any polynomial p with $p(r_k) = 1$,

$$\begin{aligned} 2^{-k}|U(f)(r_k)|^2 &= 2^{-k}|p(r_k)U(f)(r_k)|^2 \leq \|pU(f)\|^2 \\ &= \|p(M_\psi)(U(f))\|^2 = \|U(p(M_\varphi)(f))\|^2 = \|U(p_n f)\|^2 \\ &\leq \|U\|^2 \, \|p_n f\|^2 = \|U\|^2 \int_0^1 |p(x)f(x)|^2 \, dx\,. \end{aligned} \tag{1.3}$$

The Weierstrass approximation theorem implies the existence of a sequence of polynomials p_n with $p_n(r_k) = 1$, $|p_n(x)| \leq 2$ for $0 \leq x \leq 1$, and $p_n \to 0$ $d\mu$-almost-everywhere. It follows from (1.3) and the Lebesgue dominated convergence theorem that $U(f)(r_k) = 0$. Since f and k are arbitrary, the operator U is zero. In particular, the operator U is not unitary, hence, $M_\varphi \not\simeq M_\psi$.

As pointed out by Allen Shields, the assertion $M_\varphi \not\simeq M_\psi$ can be obtained more easily by observing that a rational number in $[0,1]$, say $\frac{1}{2}$, is an eigenvalue of M_ψ and is not for M_φ. The proof above suggests the following more general principle: if $\varphi(x) = x = \psi(x)$ and if μ and ν are mutually singular measures on $[0,1]$, then M_φ and M_ψ are disjoint in the sense that there are no non-zero intertwining maps between them.

<u>Example 3</u>.

Let $X = \left[-\frac{\pi}{2}, \frac{\pi}{2}\right]$, $d\mu(x) = dx$, $\varphi(x) = x$, $Y = (-\infty,\infty)$, $d\nu(x) = dx$, and $\psi(x) =$ Arctan x. Define $U : L^2(\mu) \to L^2(\nu)$ by the equation $U(f)\,(x) = (1+x^2)^{-\frac{1}{2}} f(\text{Arctan } x)$. It is then easily verified that U is unitary and that $M_\psi U = UM_\varphi$. Hence, $M_\varphi \simeq M_\psi$.

<u>Example 4</u>.

Let $X = [0,1]$, $d\mu(x) = dx$, $\varphi(x) = x$, $Y = [-1,1]$, $d\nu(x) = dx$, and $\psi(x) = |x|$. Suppose that $U : L^2(\mu) \to L^2(\nu)$ satisfies $M_\psi U = UM_\varphi$. Set $g = U(1)$ and let $h(x) = x\overline{g}(-x)$. By (1.2), for any polynomial p,

$$(1.4) \qquad \langle U(p),h\rangle = \langle U(p(M_\psi)(1)),h\rangle = \langle p(M_\varphi)(U(1)),h\rangle$$
$$= \langle p(M_\varphi)(g),h\rangle = \int_{-1}^{1} p(|x|)g(x)\overline{g}(-x)x\,dx = 0\,,$$

that is, the function h is orthogonal in $L^2(\nu)$ to $U(p)$. Since the polynomials are dense in $L^2(\mu)$, it follows that h is orthogonal in $L^2(\nu)$ to the range of U. Consequently, either $h = 0$ or the range of U is not dense in $L^2(\nu)$. In either case, the operator U is not unitary and therefore $M_\varphi \not\simeq M_\psi$.

Example 5.

Let $X = [0,1] \times [0,1]$, $d\mu(x,y) = dxdy$, $\varphi(x,y) = xy$, $Y = (-\infty,\infty)$, $d\nu(x) = dx$, and $\psi(x) = (\sin x + 1)/2$. It will be shown in Section 6 that $M_\varphi \simeq M_\psi$.

Example 6.

Let $X = [0,1]$, $d\mu(x) = dx$, $\varphi(x) = x$, $Y = [0,1]$, and $d\nu(x) = dx$. Let U_1 be the open interval of length 1/4 centered at 1/2, let U_2 be the union of the two open intervals each of length 1/16 centered at 3/16 and 13/16, and let U_3 be the union of the four open intervals each of length 1/64 centered at the midpoints of the four intervals which constitute $[0,1] \setminus (U_1 \cup U_2)$. Continue in this way to define U_k for $k \geq 1$. The set $K = [0,1] \setminus \cup \{U_k : k = 1,2,\ldots\}$ is a Cantor set with length 1/2. Let g be the function on [0,1] defined by setting g(x) equal to the distance from x to K, let α be the positive number $\alpha = \int_0^1 g(x)\,dx$ and let f be the function $\alpha^{-1}g$. The function f is continuous on [0,1], it satisfies $f(x) = 0$ for x in K, it is nonnegative, it integrates to one, and if $0 \leq a < b \leq 1$, then there is an x in (a,b) with $f(x) > 0$. The function $\psi(x) = \int_0^x f(t)\,dt$ is then a continuously differentiable function on [0,1] which is strictly increasing from $\psi(0) = 0$ to $\psi(1) = 1$. It will be shown in Section 6 that $M_\varphi \not\simeq M_\psi$.

SKETCHES OF THE EXAMPLES

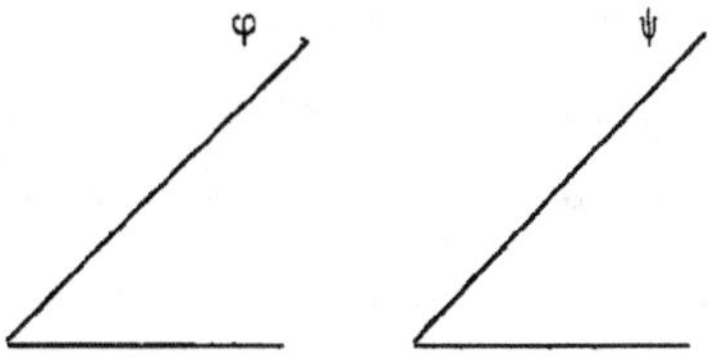

Example 1

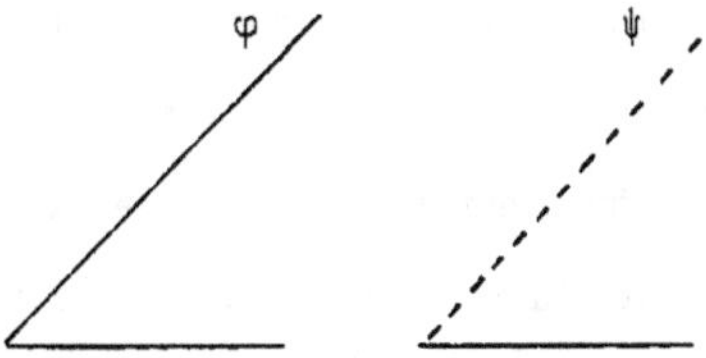

Example 2

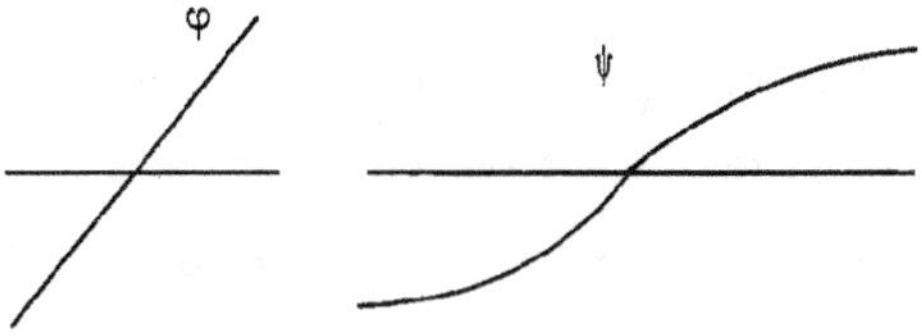

Example 3

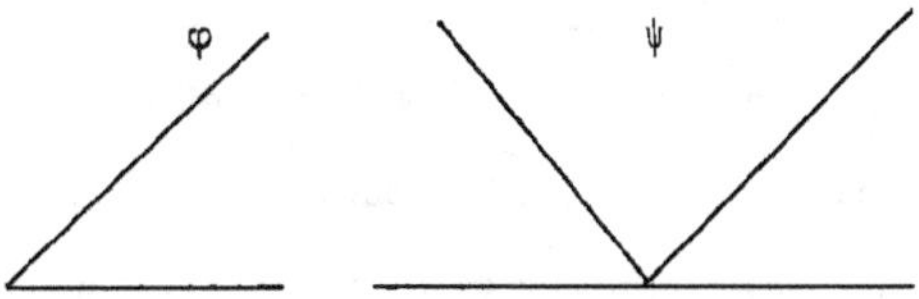

Example 4

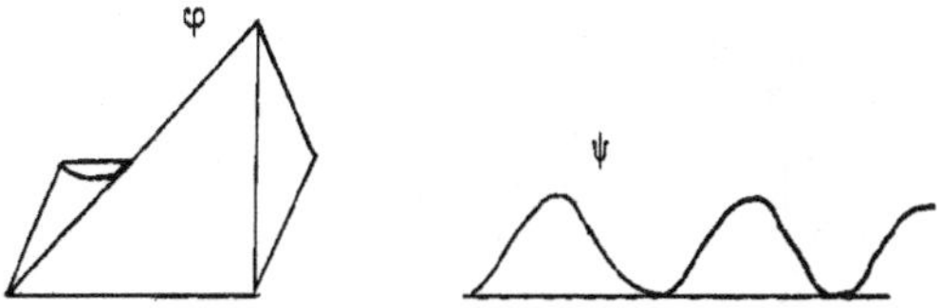

Example 5

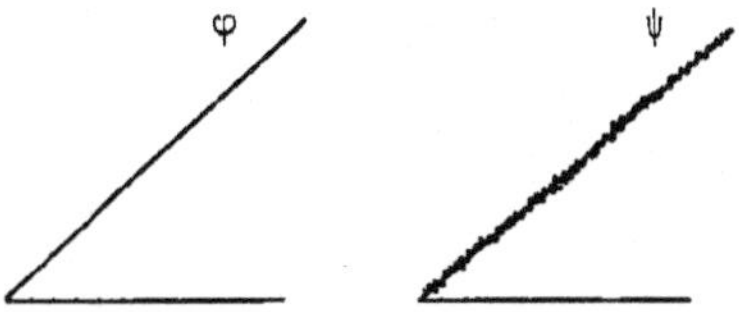

Example 6

To summarize the examples, observe that Examples 1 and 2 deal with a change of measure only; these examples suggest that a change in measure does not change the unitary equivalence class of the operator provided the two measures are mutually absolutely continuous. Examples 3, 4, and 6 deal with a change in the function. Example 3 shows that one can change the function when the new function is obtained from the old by a change of variables which preserves the measure up to mutual absolute continuity. In Example 4, the function φ is one-to-one while ψ is two-to-one and hence ψ cannot be obtained from φ by changing variables; this example suggests the general notion of multiplicity to be discussed in later sections. In Example 6, both functions are one-to-one and still one cannot change variables in a way that preserves the measure up to mutual absolute continuity. Example 5 shows that two multiplication operators which appear quite different may in fact be unitarily equivalent while Example 6 shows that two multiplication operators which appear alike may not be unitarily equivalent. These two examples also indicate that the "bare hands" approach used in Example 1 through 4 is not adequate for most problems. We turn now to more general machinery.

2. DIRECT INTEGRALS.

Let Y be a compact subset of the plane, let ν be a finite positive Borel measure on Y with closed support equal to Y, let $\mathcal{H}_y$ be a non-zero separable Hilbert space for each y in Y, and let $\mathcal{P}$ be the set of functions f from Y into $\cup \mathcal{H}_y$ such that $f(y)$ is in $\mathcal{H}_y$ for each y in Y. The reader may notice that this is exactly the kind of data required to construct a vector bundle over Y. One usually constructs a vector bundle so that the fiber spaces $\mathcal{H}_y$ fit together topologically and then one considers cross-sections f which are continuous. For direct integrals, one wants the fiber spaces to fit together measurably

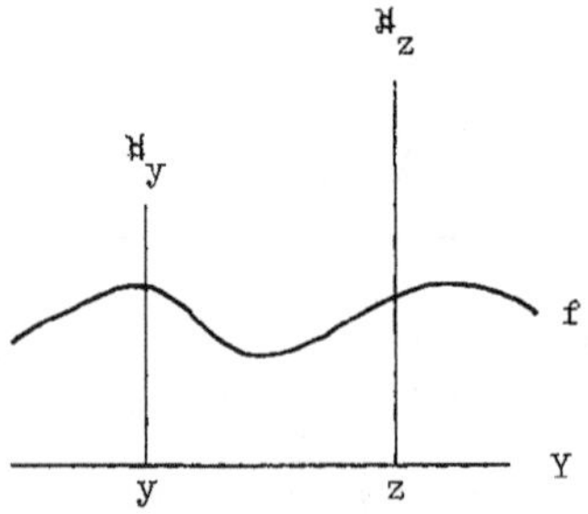

so that one can talk about measurable cross-sections. The simplest approach to the question of measurability is to select a sequence of cross-sections which behave measurably with respect to each other and span pointwise. Then one considers cross-sections which are measurable with respect to this sequence. Details follow.

Let $\mathfrak{F}$ be a countable subset of $\mathcal{P}$ such that

(2.1) for f and g in $\mathfrak{F}$, the function $y \to \langle f(y), f(y) \rangle$ is ν-measurable and

(2.2) for y in Y, the set $\{f(y) : f \text{ in } \mathfrak{F}\}$ spans $\mathcal{H}_y$.

A function g in $\mathcal{P}$ is said to be $\mathfrak{F}$-measurable if the function $y \to \langle g(y), f(y) \rangle$ is ν-measurable for each f in $\mathfrak{F}$.

A comment is in order about the existence of the set $\mathfrak{F}$. Let n be the function on Y defined by setting $n(y)$ equal to the dimension of $\mathcal{H}_y$ if this dimension is finite and equal to ∞ if the dimension of $\mathcal{H}_y$ is infinite. It follows from (2.2) and properties of Grammian matrices that $n(y)$ is equal to the rank of the matrix on $\mathfrak{F} \times \mathfrak{F}$ whose (f,g) entry is $\langle f(y), g(y) \rangle$. Hence, the existence of $\mathfrak{F}$ implies that n is ν-measurable. The converse assertion is also true, an elementary exercise for the reader.

Let $\mathcal{P}(\mathfrak{F})$ be the set of $\mathfrak{F}$-measurable functions in $\mathcal{P}$. It is immediate that $\mathcal{P}(\mathfrak{F})$ is a complex linear space and, moreover, it is a module over the ring of all ν-measurable complex functions on Y. Furthermore, if f is in $\mathcal{P}(\mathfrak{F})$, then the function $y \to \|f(y)\|$ is ν-measurable. This fact is verified in the following way. Let $\mathfrak{W}$ be the set of rational linear combinations of functions in $\mathfrak{F}$ and, for h in $\mathfrak{W}$, let χ_h be the characteristic function of the set $\{\lambda : h(\lambda) \neq 0\}$. Then, for f in $\mathcal{P}$,

$$\|f(y)\| = \sup_{h \text{ in } \mathfrak{W}} \chi_h(y) \, |\langle f(y), h(y) \rangle| \, \|h(y)\|^{-1}. \tag{2.3}$$

Hence, if f is in $\mathcal{P}(\mathfrak{F})$, then the function $y \to \|f(y)\|$ is ν-measurable. From the polarization identity for inner products, it follows that the function $y \to \langle f(y), g(y) \rangle$ is ν-measurable for every f and g in $\mathcal{P}(\mathfrak{F})$.

The direct integral $\int \oplus \mathcal{H}_y \, d\nu(y)$ is defined to be the set of equivalence classes of functions f in $\mathcal{P}(\mathfrak{F})$ such that the integral $\int \|f(y)\|^2 \, d\nu(y)$ is finite, two such functions being equivalent if they agree $d\nu$-almost-everywhere. An inner product on $\int \oplus \mathcal{H}_y \, d\nu(y)$ is defined by the equation $\langle f, g \rangle = \int \langle f(y), g(y) \rangle \, d\nu(y)$ and the usual proof of completeness in L^p-spaces shows that $\int \oplus \mathcal{H}_y \, d\nu(y)$ is complete, an exercise for the reader. Thus, the direct integral $\int \oplus \mathcal{H}_y \, d\nu(y)$

is a Hilbert space.

Let $S(Y,\nu,n)$ be the operator on $\int \oplus \mathcal{H}_y \, d\nu(y)$ defined by the equation

$$S(Y,\nu,n)(f)(y) = yf(y). \tag{2.4}$$

The importance of this operator for this paper is that the triple (Y,ν,n) is a complete unitary invariant for the operator. More precisely, let (Y',ν',n') be a second triple of the same type.

THEOREM 1. The operator $S(Y,\nu,n)$ is unitarily equivalent to the operator $S(Y',\nu',n)$ if and only if $Y = Y'$, ν is mutually absolutely continuous with respect to ν', and $n(y) = n'(y)$ $d\nu$-almost-everywhere.

The reader is referred to Dixmier [4, Chapter II] for a proof of Theorem 1. It is hoped that the concrete computations in Section 1 serve to make the theorem plausible.

Notice that the set $\mathfrak{F}$ is not referred to in the notation for the direct integral $\int \oplus \mathcal{H}_y \, d\nu(y)$ or for the operator $S(Y,\nu,n)$. The reason for this is Theorem 1; if a second set $\mathfrak{F}'$ is used, then the resulting operator is unitarily equivalent to the original.

A comment is in order about the relationship between Theorem 1 and the theory of normal operators on a Hilbert space. An operator N is said to be normal if $N^*N = NN^*$. The spectral theorem states that the following four assertions are equivalent.

A. The operator N is a normal operator on a separable Hilbert space.

B. The operator N is unitarily equivalent to a multiplication operator on a separable L^2 space.

C. The operator N is unitarily equivalent to $S(Y,\nu,n)$ for some (Y,ν,n).

D. There is a spectral measure E such that N is unitarily equivalent to $\int \lambda \, dE(\lambda)$.

The three representations for normal operators given in B, C, and D have various uses. Theorem 1 shows that the representation in terms of direct integrals (C) exhibits unitary invariants for the operator. These unitary invariants Y, ν, and n for the normal operator N are referred to in the following way: the set Y is the spectrum of N, the measure ν is the scalar spectral measure of N, and the function n is the multiplicity function of N.

In this paper, a direct integral for a multiplication operator is produced. Thus, it is proved that B implies C. However, the interest is not with the implication B implies C which could be obtained, perhaps more cheaply, by showing that $B \Rightarrow A \Rightarrow D \Rightarrow C$. The objective is to obtain the direct integral for M_φ in a way that one can compute the unitary invariants Y, n, and ν in terms of φ and μ. One way to do this is to use the theorem on disintegration of measures.

3. DISINTEGRATION OF MEASURES.

Let X be a locally compact separable metric space and let P(X) be the set of regular Borel probability measures on X. Fix μ in P(X). Let Φ be a μ-essentially bounded Borel measurable function on X, let Y be the μ-essential range of Φ, and let ν in P(Y) be the measure $\nu(E) = \mu(\Phi^{-1}(E))$. A useful example to keep in mind occurs when $X = [0,1] \times [0,1]$, $\Phi(u,v) = u$, and $Y = [0,1]$. One then has $\nu(E) = \mu(E \times [0,1])$.

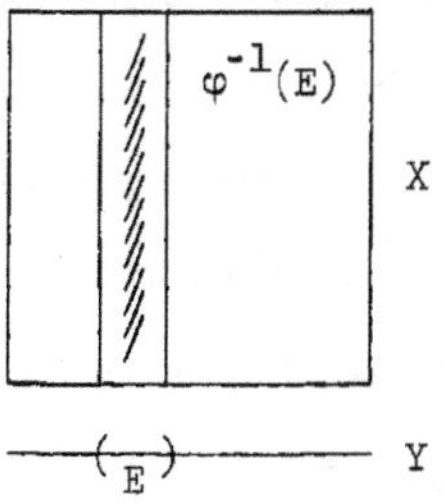

A disintegration of μ with respect to Φ is a function $y \to \mu_y$ from Y into P(X) such that

(3.1) $\mu_y(X/\Phi^{-1}(\{y\})) = 0$ $d\nu$-almost-everywhere and

(3.2) for every Borel set E of X, the function $y \to \mu_y(E)$ is ν-measurable and $\mu(E) = \int \mu_y(E)\, d\nu(y)$.

The following theorem establishes the existence and uniqueness of a disintegration. It also shows that the disintegration depends only on the equivalence class of Φ in $L^\infty(\mu)$. Thus, it makes sense to refer to a disintegration of μ with respect to an equivalence class φ in $L^\infty(\mu)$ and in this case condition (3.1) holds for any Borel representative Φ of φ. This theorem is proved in many places in various forms [1, Theorm 2; 2, Proposition 1, 5.3; 8, Chapter V,

Section 8].

THEOREM 2. There is a disintegration of μ with respect to Φ. If $y \to \mu_y$ and $y \to \mu'_y$ are disintegrations of μ with respect to Φ, then $\mu_y = \mu'_y$ $d\nu$-almost-everywhere. If $\Phi = \Psi$ $d\mu$-almost-everywhere and if $y \to \mu_y$ is a disintegration of μ with respect to Φ, then $y \to \mu_y$ is a disintegration of μ with respect to Ψ.

If $X = [0,1] \times [0,1]$, $\varphi(u,v) = u$, and μ is planar Lebesgue measure, then $d\nu(u) = du$ and a disintegration of μ with respect to φ is obtained by setting μ_u equal to linear Lebesgue measure on the set $\{u\} \times [0,1]$. Thus, in this special case, Theorem 2 is Fubini's theorem. If $X = [0,1] \times [0,1]$, $\varphi(u,v) = u$, and μ is linear Lebesgue measure on $\{(u,u) : 0 \leq u \leq 1\}$ times the normalization constant $2^{-1/2}$, then $d\nu(u) = du$ and a disintegration of μ with respect to φ is obtained by setting μ_u equal to a unit point mass at (u,u). A third elementary example is obtained by taking $X = \{z : |z| \leq 1\}$, $\varphi(z) = |z|$, and μ equal to planar Lebesgue measure divided by π. In this case $Y = [0,1]$ and $d\nu(r) = 2r\,dr$. A disintegration is obtained by setting μ_r equal to $(2\pi r)^{-1}$ times linear Lebesgue measure on $|z| = r$; here, the disintegration theorem yields integration in polar coordinates.

Example 5 of Section 1 provides two non-trivial examples of the disintegration theorem. Let $X = [0,1] \times [0,1]$, $d\mu(u,v) = dudv$, and $\varphi(u,v) = uv$. Then $Y = [0,1]$ and integration shows that $\nu([0,y]) = y - y \log y$, hence,

$$d\nu(y) = \log \frac{1}{y}\, dy\,. \tag{3.3}$$

The disintegration theorem asserts the existence of a measure μ_y supported on the intersection of the hyperbola $uv = y$ with X such that $\mu(E) = \int \mu_y(E) \log \frac{1}{y}\, dy$ for all Borel sets E in X.

The other function in Example 5 of Section 1 is the function $\psi(x) = (\sin x + 1)/2$ defined on $X = (-\infty,\infty)$. Here, it will be convenient to replace linear Lebesgue measure on X by a mutually absolutely continuous probability measure μ on X; as Example 1 indicates, such a change does not change the unitary equivalence class of the multiplication operator. It is convenient to define

$$(3.4) \qquad d\mu(x) = \left[\frac{1}{2\pi} \chi_0(x) + \frac{1}{4\pi} \sum_{|n|>0} 2^{-|n|} \chi_n(x) \right] dx$$

where χ_n is the characteristic function of the interval $\left[n\pi - \frac{\pi}{2}, n\pi + \frac{\pi}{2} \right)$. The essential range of ψ is the set $Y = [0,1]$ and, for y in Y, a computation gives $\nu([0,y]) = \frac{\pi}{2} + \sin^{-1}(2y - 1)$, hence,

$$(3.5) \qquad d\nu(y) = (y - y^2)^{-1/2} \, dy .$$

The disintegration theorem produces a measure μ_y supported on the infinite discrete set $\psi^{-1}(y)$ such that $\mu(E) = \int_0^1 \mu_y(E)(y - y^2)^{-1/2} \, dy$ for any Borel set E in $(-\infty, \infty)$.

There is a connection between disintegrations and expectation operators which will be useful in Section 5. Let $\mathcal{L}^1(X,\mu)$ be the set of all μ-summable Borel function on X and let $L^1(\nu)$ be the usual Lebesgue space of equivalence classes of functions in $\mathcal{L}^1(Y,\nu)$. There is an expectation operator E from $\mathcal{L}^1(X,\mu)$ into $L^1(\nu)$ defined by the equation

$$(3.6) \qquad \int \psi E(f) d\nu = \int (\psi \circ \varphi) f \, d\mu$$

for all ψ in $L^\infty(\nu)$. That this equation does define an operator is an elementary consequence of the Radon-Nikodym Theorem. The connection between expectations and disintegrations is contained in the following theorem.

THEOREM 3. The function $y \to \mu_y$ is a disintegration of μ with respect to φ if and only if

$$(3.7) \qquad E(f)(y) = \int f \, d\mu_y \quad d\nu \text{ — a.e.}$$

for each f in $\mathcal{L}^1(x,\mu)$.

The proof of this theorem is elementary and can be found in [1]; in [1], the proof of Theorem 2 makes use of Theorem 3. An informal statement of Theorem 3 is that a disintegration of μ with respect to φ evaluates the expectation operator E pointwise.

4. THE DIRECT INTEGRAL FOR A MULTIPLICATION OPERATOR.

As in Section 3, let X be a locally compact separable metric space, let μ be a Borel probability measure on X, let φ be in $L^{\infty}(\mu)$, and let $\to \mu_y$ be a disintegration of μ with respect to φ. The purpose of this section is to establish the following theorem.

THEOREM 4. The operator M_{φ} is unitarily equivalent to $S(Y,\nu,n)$ on a direct integral $\int \oplus \mathcal{H}_y \, d\nu(y)$ where the set Y is the essential range of φ, the measure ν is $\mu \circ \varphi^{-1}$, and the Hilbert space $\mathcal{H}_y$ is $L^2(\mu_Y)$.

To prove this theorem, one must first produce a countable set $\mathfrak{F}$ of functions $f : Y \to \bigcup \mathcal{H}_y$ satisfying conditions (2.1) and (2.2). Then, having constructed the direct integral $\int \oplus \mathcal{H}_y \, d\nu(y)$ with respect to the set $\mathfrak{F}$, one must produce a unitary operator U from $L^2(\mu)$ onto $\int \oplus \mathcal{H}_y \, d\nu(y)$ such that $UM_{\varphi} = S(Y,\nu,n)U$. Before defining $\mathfrak{F}$ and U, four preliminary observations are needed.

Let $C_0(X)$ be the space of continuous complex functions on X with compact support. The topological assumptions on X guarantee the existence of a countable set $\mathfrak{D}$ in $C_0(X)$ such that any function in $C_0(X)$ is a uniform limit of a sequence of functions in $\mathfrak{D}$. This countable set $\mathfrak{D}$ has the following property.

(4.1) If τ is any finite Borel measure on X, then $\mathfrak{D}$ is dense in $L^2(\tau)$.

The reason for this is that any finite Borel measure on a separable locally compact metric space is regular [9, Theorem 2.18].

Let $\mathfrak{L}^{\infty}(X)$ be the set of all bounded Borel measurable functions on X. For f in $\mathfrak{L}^{\infty}(X)$, define $\hat{f} : Y \to \bigcup L^2(\mu_y)$ by setting $\hat{f}(y)$ equal to the equivalence class of f in $L^2(\mu_y)$. Let f and g be in $\mathfrak{L}^{\infty}(X)$ and let Φ in $\mathfrak{L}^{\infty}(X)$ be a representative of φ.

(4.2) $$\langle \hat{f}(y), \hat{g}(y) \rangle = \int f\bar{g} \, d\mu_y = E(\bar{g})(y) .$$

$d\nu$-almost-everywhere.

(4.3) $$\int \langle \hat{f}(y), \hat{g}(y) \rangle \, d\nu(y) = \int f\bar{g} \, d\mu .$$

(4.4) If h is a continuous complex function on the plane, then $(h\circ\varphi)f\hat{}\,(y) = h(y)\,\hat{f}(y)$ $d\nu$-almost-everywhere.

Equation (4.2) follows from Theorem 3, (4.3) follows from (4.2) and (3.6) with $\psi \equiv 1$, and (4.4) is an immediate consequence of (3.1).

To prove Theorem 4, define $\mathfrak{F}$ to be the set $\{\hat{f} : f \text{ in } \mathfrak{D}\}$. That $\mathfrak{F}$ satisfies conditions (2.1) and (2.2) follows immediately from (4.1) and (4.2). One now constructs the direct integral $\int \oplus \mathcal{H}_y \, d\nu(y)$ with respect to the set $\mathfrak{F}$. Equations (4.2) and (4.3) show that this direct integral contains the set $\{\hat{f} : f \text{ in } \mathfrak{L}^\infty(X)\}$. Moreover, (4.3) guarantees the existence of an isometry U from $L^2(\mu)$ into $\int \oplus \mathcal{H}_y \, d\nu(y)$ such that $U([f]) = \hat{f}$ for all f in $\mathfrak{L}^\infty(X)$. Here, $[f]$ denotes the equivalence class of f in $L^2(\mu)$. To show that U is onto (hence unitary), suppose that g in $\int \oplus \mathcal{H}_y \, d\nu(y)$ is orthogonal to the range of U. Then for all f in $\mathfrak{D}$ and for all h continuous on the complex plane, by (4.4), $0 = \langle U([(h\circ\varphi)f]),g\rangle = \int h(y)\langle \hat{f}(y),g(y)\rangle \, d\nu(y)$. It follows that $\langle \hat{f}(y),g(y)\rangle = 0$ for every f in $\mathfrak{D}$ and every y in $Y\setminus F$. It follows from (4.1) that $g(y) = 0$ for y in $Y\setminus F$ and thus $g = 0$. Hence, U is onto. Finally, that $UM_\varphi = S(Y,\nu,n)U$ is a consequence of (4.4) with $h(z) = z$. This completes the proof of Theorem 4.

5. THE ESSENTIAL PRE-IMAGE.

The direct integral $\int \oplus \mathcal{H}_y \, d\nu(y)$ for M_φ produced in Section 4 is a complete unitary invariant for M_φ as indicated by Theorem 1. This invariant is determined by the triple (Y,ν,n) where Y is the essential range of φ and ν is the measure $\mu\circ\varphi^{-1}$. The multiplicity function n is defined by setting $n(y)$ equal to the dimension of $L^2(\mu_y)$ where $y \to \mu_y$ is a disintegration of μ with respect to φ. It is an elementary verification that the dimension of $L^2(\mu_y)$ is equal to the number of points in the closed support of μ_y if the closed support is finite. Otherwise, the space $L^2(\mu_y)$ is infinite dimensional. Thus, if support (μ_y) denotes the closed suport of μ_y, then

(5.1) $$n(y) = \#\ \text{support}\ (\mu_y)$$

where $\#A$ is the cardinality of the set A when A is finite and $\#A$ is ∞ when A is infinite.

To use Equation (5.1) for computing the multiplicity function of M_φ, one must produce explicitly the disintegration of μ with respect to φ. In most cases, this is difficult; if the reader is unconvinced of this, I recommend that he or she attempt to work out the disintegrations for the two functions in Example 5. In this section, we develop a way to compute the number of points in support (μ_y) without actually producing the measure μ_y. This computation makes use of the notion of essential pre-image due to T. L. Kriete [7].

Let X, μ, φ, and Y be as in Sections 3 and 4, let U be an open subset of X, and let y be a point in Y. Define

$$D_U(y) = \liminf_{\delta \downarrow 0} \frac{\mu(\varphi^{-1}(B_\delta(y) \cap U)}{\mu(\varphi^{-1}(B_\delta(y)))} \tag{5.2}$$

where $B_\delta(y)$ is the open disk of radius δ about y. Taking limits formally in (5.2) gives

$$D_U(y) = \frac{\mu(\{x \text{ in } U : \varphi(x) = y\})}{\mu(\{x \text{ in } X : \varphi(x) = y\})} . \tag{5.3}$$

Equation (5.3) is valid only when the denominator on the right side is non-zero and this will usually not be the case; notice that the denominator on the right side of (5.2) is always non-zero because y is in the essential range of φ. Equation (5.3) suggests the following interpretation of $D_U(y)$: the number $D_U(y)$ is equal to or greater than the probability that a solution to the equation $\varphi(x) = y$ lies in U.

The essential pre-image of φ at y, denoted $\varphi_\mu^{-1}(y)$, is the set of points x in X such that $D_U(y) > 0$ for every open set U containing x. Informally, the set $\varphi_\mu^{-1}(y)$ consists of those points x such that, for every open set U containing x, there is a positive probability that a solution to the equation $\varphi(u) = y$ lies in U.

It is elementary to verify that for a continuous function φ, the essential pre-image of φ at y is contained in the pre-image of φ at y. However, this containment can be strict even for all points y in a set of positive ν-measure. I refer the reader to [1] for these arguments. Here, we look at an elementary example that indicates how the essential pre-image can be strictly contained in the pre-image and then we prove that the essential pre-image can be used to compute the multiplicity function for M_φ.

Let X be the unit interval $[0,1]$, let $d\mu(x) = dx$, and define φ on X by setting $\varphi(x) = 72\left(x - \frac{1}{3}\right)^3$ for $0 \leq x \leq \frac{1}{2}$ and $\varphi(x) = \frac{2}{3} - x$ for $\frac{1}{2} \leq x \leq 1$. then $\varphi^{-1}(0) = \left\{\frac{1}{3}, \frac{2}{3}\right\}$. However, if $U = \left(0, \frac{1}{2}\right)$ and $0 < \delta < \frac{1}{3}$, then $\mu(\varphi^{-1}(B_\delta(0) \cap U) = 2\left(\frac{\delta}{72}\right)^{1/3}$ and $\mu(\varphi^{-1}(B_\delta(0)) = 2\left(\frac{\delta}{72}\right)^{1/3} + 2\delta$. It follows that $D_U(0) = 0$ and thus $\frac{1}{3}$ is not in $\varphi_\mu^{-1}(0)$. Thus, $\varphi_\mu^{-1}(0) = \left\{\frac{2}{3}\right\} \subsetneq \left\{\frac{1}{3}, \frac{2}{3}\right\} = \varphi^{-1}(0)$.

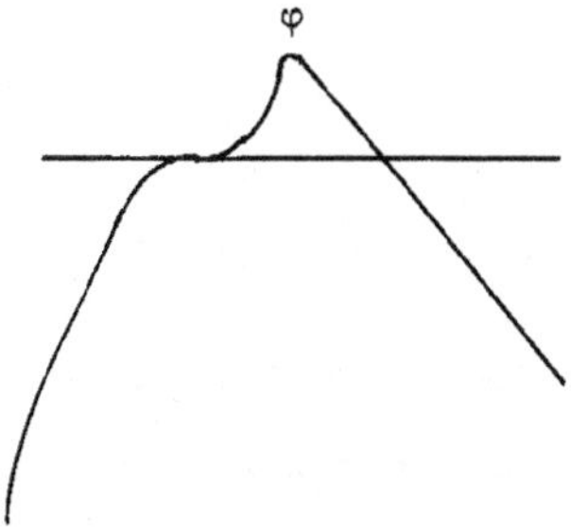

THEOREM 5. The multiplicity function n for M_φ satisfies the equation

$$n(y) = \#\varphi_\mu^{-1}(y) \tag{5.4}$$

$d\nu$-almost-everywhere.

The proof of Theorem 5 makes use of the following assertion due to Besicovitch [3]: if f is a function in $L^1(\nu)$, then

$$f(y) = \lim_{\delta \downarrow 0} \frac{1}{\nu(B_\delta(y))} \int_{B_\delta(y)} f \, d\nu \tag{5.5}$$

$d\nu$-almost-everywhere. This fact can be viewed as a generalization of the fundamental theorem of calculus.

To prove Theorem 5, let $\mathfrak{U}$ be a countable basis for the topology on X, let U be in $\mathfrak{U}$, and let χ_U be the characteristic function of U. By (3.7),

$$E(\chi_U) = \int \chi_U \, d\mu_y = \mu_y(U) \tag{5.6}$$

$d\nu$-almost-everywhere and by (5.5) and (3.6) with $\psi \equiv 1$,

(5.7) $$E(X_U)(y) = \lim_{\delta \downarrow 0} \frac{1}{\nu(B_\delta(y))} \int_{B_\delta(y)} E(X_U)\, d\nu = D_U(y)\,.$$

$d\nu$-almost-everywhere. Equating (5.6) with (5.7) and using the fact that $\mathfrak{U}$ is countable, one obtains a Borel set $\Omega \subset Y$ such that $\nu(Y \setminus \Omega) = 0$ and

(5.8) $$\mu_y(U) = D_U(y)$$

for each y in Ω and U in $\mathfrak{U}$. It follows immediately from (5.8) that, for y in Ω,

(5.9) $$\varphi_\mu^{-1}(y) = \text{support } (\mu_y)\,.$$

The theorem is a consequence of (5.9) and (5.1).

6. EXAMPLES REVISITED.

We return now to the six examples of Section 1. For each of these examples let (Y,ν,n) and (Y',ν',n') be the unitary invariants of Section 3 for M_φ and M_ψ respectively. This means that the symbols Y and ν are used twice, but I do not think this will cause confusion. In each example, the sets Y and Y' are equal. Therefore, the unitary equivalence problem involves comparing ν with ν' and n with n'.

In Example 1, it is easily verified that ν is linear Lebesgue measure, that ν' is mutually absolutely continuous with respect to ν, that $n \equiv 1$, and that $n' \equiv 1$. Hence, $M_\varphi \simeq M_\psi$ as observed previously. In Example 2, the measure ν' is not absolutely continuous with respect to ν and therefore $M_\varphi \not\simeq M_\psi$.

To apply the machinery of Sections 4 and 5 to Example 3, one must replace linear Lebesgue measure on $\left[-\frac{\pi}{2}, \frac{\pi}{2}\right]$ by a convenient mutually absolutely continuous probability measure and likewise for linear Lebesgue measure on $(-\infty,\infty)$. We use $\pi^{-1}\,dx$ for the former and $\pi^{-1}(1+x^2)^{-1}\,dx$ for the latter. Then

(6.1) $$\nu'([0,y]) = \pi^{-1} \int_{-\infty}^{\tan y} (1+x^2)^{-1}\,dx$$

and therefore $d\nu'(y) = \pi^{-1}dy = d\nu(y)$. Also, $n \equiv 1$ and $n' \equiv 1$ so that

$M_\varphi \simeq M_\psi$ as asserted.

In Example 4, it is immediate that ν is linear Lebesgue measure on $[0,1]$ and that $n \equiv 1$. After normalizing linear Lebesgue measure on $[-1,1]$ to be a probability measure, one obtains $\nu' = \nu$. Finally, it is easily verified that the essential pre-image of the function $\psi(x) = |x|$ is equal to its pre-image at each point y in $[0,1]$. Hence, $n' \equiv 2$ by Theorem 5 and therefore $M_\varphi \not\simeq M_\psi$. The assertion $n' \equiv 2$ can also be obtained by using Equation (5.1) because, in this case, it is not difficult to write down a disintegration explicitly. In fact, one can set $\mu_y = \frac{1}{2}\delta_y + \frac{1}{2}\delta_{-y}$ where δ_z is the unit point mass at z. Thus, by Equation (5.1), $n'(y) = \#\text{ support }(\mu_y) = \#\{y, -y\}$ and therefore $n'(y) = 2$ for all $0 < y \leq 1$.

For Example 5, Equation (3.3) says that $d\nu(y) = \log \frac{1}{y}\, dy$ and, after a mutually absolutely continuous change in measure given by Equation (3.4), Equation (3.5) asserts that $d\nu'(y) = (y - y^2)^{-1/2}\, dy$. Hence, the measures ν and ν' are mutually absolutely continuous. To compute $\varphi_\mu^{-1}(y)$, take y in $(0,1)$ and (s,t) in $(0,1) \times (0,1)$ with $y = \varphi(s,t) = st$. Let U be an open set containing (s,t). There is then an open rectangle $R = \{(u,v) : a < u < b,\ c < v < d\}$ and an $\varepsilon > 0$ such that (i) $\varepsilon < y$, (ii) $(s,t) \in R \subset U$, and (iii) the set $\varphi^{-1}(B_\varepsilon(y)) = \{(u,v) : y - \varepsilon < uv < y + \varepsilon\}$ intersects neither of the two horizontal line segments which are contained in the boundary of R. Direct computation shows that for $0 < \delta \leq \varepsilon$

(6.2) $$\mu\left(\varphi^{-1}(B_\delta(y)\right) = \int_{y-\delta}^{y+\delta} \log \frac{1}{t}\ dt \leq 2\delta \log\left(\frac{1}{y - \varepsilon}\right)$$

and

(6.3) $$\mu\left(\varphi^{-1}(B_\delta(y)) \cap R\right) = \int_a^b \frac{y + \delta}{u} - \frac{y - \delta}{u}\ dy = 2\delta \log \frac{b}{a}.$$

By (6.2) and (6.3),

(6.4) $$\frac{\mu\left(\varphi^{-1}(B_\delta(y)) \cap U\right)}{\mu\left(\varphi^{-1}(B_\delta(y))\right)} \geq \frac{\log \frac{b}{a}}{\log\left(\frac{1}{y - \varepsilon}\right)}$$

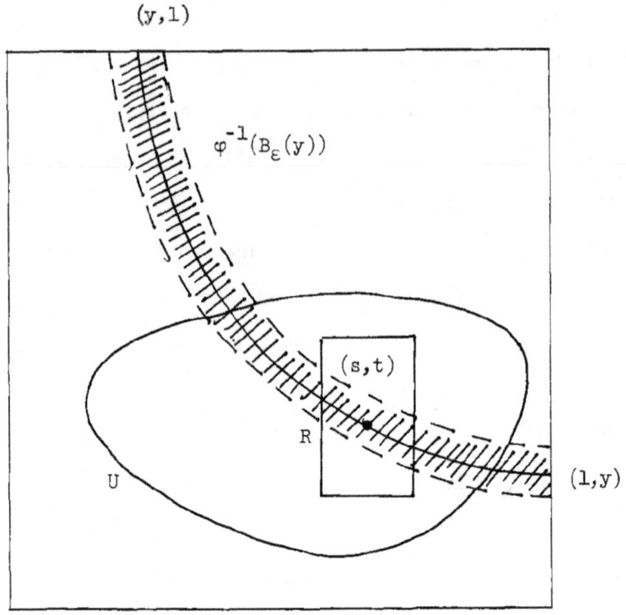

for all $0 < \delta \leq \varepsilon$. Equations (6.4) and (5.2) imply that $D_U(s,t) > 0$ and therefore (s,t) is in $\varphi_\mu^{-1}(y)$. This proves that for y in $(0,1)$

$$\varphi^{-1}(y) \cap ((0,1) \times (0,1)) \subset \varphi_\mu^{-1}(y) \tag{6.5}$$

and thus, by (5.4), one has $n(y) \equiv \infty$ $d\nu$-almost-everywhere.

To compute n', fix y in $(0,1)$ and s in $(-\infty,\infty)$ such that $y = \varphi(s) = \frac{\sin s + 1}{2}$. Suppose that s is in $\left(k\pi - \frac{\pi}{2},\ k\pi + \frac{\pi}{2}\right)$. Let U be an open set containing s and let $\varepsilon > 0$ be chosen so that, for each $0 < \delta \leq \varepsilon$, the set $\varphi^{-1}(B_\delta(y))$ is the union of open intervals G_n^δ contained in $n\pi - \frac{\pi}{2},\ n\pi + \frac{\pi}{2}$. Assume further that ε is small enough so that G_k^ε is contained in U. Then, for $0 < \delta \leq \varepsilon$,

$$\frac{\mu\left(\varphi^{-1}(B_\delta(y))U\right)}{\mu\left(\varphi^{-1}(B_\delta(y))\right)} \geq \frac{\mu(G_k^\delta)}{\Sigma\, \mu(G_n^\delta)} \tag{6.6}$$

$$= \begin{cases} \frac{1}{2} & \text{if } k = 0 \\ \frac{1}{2^{|k|+1}} & \text{if } k \neq 0\,. \end{cases}$$

Hence, by equation (5.2), $D_U(s) > 0$. This proves that $\varphi^{-1}(y) = \varphi_\mu^{-1}(y)$ for all y in $(0,1)$ and therefore $n'(y) = \infty$ $d\nu$-almost-everywhere by Equation (5.4). Since the measures ν and ν' are mutually absolutely continuous and $n(y) = \infty = n'(y)$ $d\nu$-almost-everywhere, Theorem 5 implies that $M_\varphi \simeq M_\psi$.

In Example 6, the measure ν is linear Lebesgue measure on $[0,1]$. However,

$$(6.7) \qquad 1 - \nu(\psi(K)) = \nu(\psi([0,1]\setminus K)) = \nu\left(\psi\left(\bigcup_{n=1}^{\infty} U_n\right)\right)$$

$$= \sum_{n=1}^{\infty} \int_{U_n} f(x)dx = \int_0^1 f(x)dx = 1$$

and thus $\nu(\psi(K)) = 0$. But $\nu'(\psi(K)) = \nu(\psi^{-1}(\psi(K)) = \nu(K) = 1/2$. This proves that ν and ν' are not mutually absolutely continuous and therefore $M_\varphi \not\simeq M_\psi$.

7. OBSERVATIONS.

Let (Y,ν,n) be the unitary invariant associated with the multiplication operator M_φ on $L^2(\mu)$ and suppose that φ is continuous. Examples such as $\varphi(x) = |x|$ on $L^2([-1,1])$ lead one to hope that the multiplicity function n at y can be computed by counting the number of points in $\varphi^{-1}(y)$. In fact, one has only the inequality $n(y) \le \#\varphi^{-1}(y)$ in general. Theorem 5 indicates for the case when μ is linear Lebesgue measure on the unit interval and φ is real-valued. The general idea of these results is that $\varphi_\mu^{-1}(y) = \varphi^{-1}(y)$ when the set $\{x : \varphi'(x) = 0\}$ is not too big, that is, the function φ is not flat too often. Example 6 indicates that a smoothness condition such as C^1 is not sufficient.

Finally, the development in this paper is carried out under the assumption that the operator M_φ acts on the space $L^2(\mu)$ where μ is a Borel measure on a locally compact separable metric space X. There is a method for computing the invariant (Y,ν,n) when the only assumption is that $L^2(\mu)$ is separable [1, Theorem 9]. In this situation, the notion of essential pre-image breaks down. However, one is able to compute the multiplicity function n using the quantities $D_U(y)$ defined by Equation (5.2) where U is restricted to a countable dense subset of the measure algebra of μ. The reader is referred to [1] for details.

REFERENCES

[1] M. B. Abrahamse and T. L. Kriete, The spectral multiplicity of a multiplication operator, Indiana U. Math. J. 22 (1973), 845-857.

[2] W. Arveson, Operator algebras and invariant subspaces, Ann. Math. 100 (1974), 433-532.

[3] A. S. Besicovitch, A general form of the covering principle and relative differentiation of additive functions II, Proc. Cambridge, Phil. Soc. 42 (1946), 1-10.

[4] J. Dixmier, Les algebras d'operateurs dans l'espace Hilbertien, Gauthier-Villars, Paris, 1969.

[5] P. R. Halmos, Measure Theory, Van Nostrand, New York, 1950.

[6] P. R. Halmos, A Hilbert Space Problem Book, Van Nostrand, Princeton, N.J., 1967.

[7] T. L. Kriete, Complete non-self-adjointness of almost self-adjoint operators, Pacific J. Math. 42 (1972), 413-437.

[8] K. R. Parthasarathy, Probability Measures on Metric Spaces, Academic Press, New York and London, 1957.

[9] W. Rudin, Real and Complex Analysis, McGraw-Hill, New York, 1966.

UNIVERSITY OF VIRGINIA

AND

LOYOLA UNIVERSITY

COMPOSITION OPERATORS ON HILBERT SPACES

Eric A. Nordgren

INTRODUCTION

Let X be a set and suppose V is a vector space of complex valued functions on X under the pointwise operations of addition and scalar multiplication. If T is a mapping of X into X such that the composite $f \circ T$ of f with T is in V whenever f is, then T induces a linear transformation C_T on V that sends f into $f \circ T$.

We are interested in the case in which V is a Hilbert space and C_T turns out to be a bounded operator, which we call a composition operator. There are two contexts in which such a study can be carried out. In one V is the L^2 space of a measure and in the other V is a functional Hilbert space such as H^2 of the disc. The first situation arose in B. O. Koopman's 1932 formulation of classical mechanics [Kp]. The second has appeared, at least implicitly, in connection with Markov processes (see [KM]), and it has been the subject of considerable explicit study. Part I of these notes will be devoted to the L^2 context, and Part II will be devoted to functional Hilbert spaces, particularly H^2 of the disc.

Composition operators can be studied in many contexts other than ours. It is easy to see that much of what is done in these notes remains true if 2 is simply replaced by p and appropriate minor changes are made. Indeed many of the references quoted are written in terms of L^p or H^p. Yet other contexts have proven to be of interest as well (see e.g., [M2], [Km1], [KS], [KM], [Rn1], [Rn2], [Sw], [T2]).

These notes constitute an expanded version of three lectures in which I tried to survey some of the theory of composition operators as developed in the last twelve years or so. Aside from organization there is little that is new here. The one claim to novelty is the use of the harmonic majorant technique to prove boundedness and estimate norms in the H^2 case. Its use stems from the fundamental papers of Rudin, [Rd1] and [Rd2]. I became aware of the possibility of using it shortly after my own paper on the subject was written. It has since been used in Swanton's thesis

[Sw] to study composition operators on multiply connected domains.

Part I. L^2 spaces

A. Boundedness. Let $(X,\mathcal{S},m)$ be a measure space. To avoid all measure theoretic pathology we will assume X is a standard Borel space, i.e., X is isomorphic to a Borel subset of a complete separable metric space and $\mathcal{S}$ is the σ-algebra of Borel subsets of X. The measure m will be finite or σ-finite.

It is useful to use a somewhat more general definition of composition operator than that given in the introduction. Let Y be a measurable subset of X, and let T be a measurable transformation of Y into X, i.e., T is a function from Y to X such that $T^{-1}E \in \mathcal{S}$ whenever $E \in \mathcal{S}$. Define C_T on the measurable complex valued functions on X by

(1)
$$C_Tf(x) = \begin{cases} f(T(x)) & \text{if } x \in Y \\ 0 & \text{if } x \in X\setminus Y. \end{cases}$$

Our first concern is the question: under what circumstances does C_T induce an operator on $L^2(m)$? It is required that C_Tf depend only on the equivalence class of f in $L^2(m)$, or what amounts to the same thing, that $C_Tf = 0$ a.e. whenever $f = 0$ a.e. Further, it is required that for some constant M and for all f in $L^2(m)$, we have $\|C_Tf\| \le M\,\|f\|$, where $\|f\|$ is the usual norm in $L^2(m)$, $\|f\| = (\int |f|^2\, dm)^{1/2}$. Two necessary conditions are easily obtained from these requirements. If $E \in \mathcal{S}$ and $m(E) = 0$, then $\chi_E = 0$ in $L^2(m)$. Since $T^{-1}E \subset Y$, we have

$$\chi_{T^{-1}E} = C_T\chi_E = 0,$$

or equivalently $m(T^{-1}E) = 0$. Thus the measure mT^{-1}, defined by $mT^{-1}(F) = m(T^{-1}F)$ for F in $\mathcal{S}$, is absolutely continuous with respect to m, which is the first condition. In this case we write $mT^{-1} \ll m$, as usual. The second condition is a consequence of the transformation of integral formula

(2)
$$\int_Y f\circ T\, dm = \int_X f\, dmT^{-1}$$

(see [H2], page 163). Write the norm inequality $\|C_T f\|^2 \leq \|C_T\|^2 \|f\|^2$ in terms of integrals, and transform the left hand side to obtain

$$(3) \qquad \int |f|^2 \, dmT^{-1} \leq \|C_T\|^2 \int |f|^2 \, dm .$$

Since $mT^{-1} \ll m$ the left side of the last inequality can be rewritten in terms of the Radon-Nikodym derivative dmT^{-1}/dm of mT^{-1} with respect to m to produce

$$(4) \qquad \int |f|^2 \, \frac{dmT^{-1}}{dm} \, dm \leq \|C_T\|^2 \int |f|^2 \, dm .$$

As this inequality holds for all f in $L^2(m)$, it follows that dmT^{-1}/dm is essentially bounded; in fact

$$(5) \qquad \frac{dmT^{-1}}{dm} \leq \|C_T\|^2 \quad \text{a.e.}$$

We have now obtained two necessary conditions for C_T to be an operator on $L^2(m)$, and we have established one half of each of the following assertions.

THEOREM 1. Necessary and sufficient conditions for a measurable transformation T to induce a bounded operator on $L^2(m)$ defined by (1) are $mT^{-1} \ll m$ and dmT^{-1}/dm is bounded. In this case

$$\|C_T\| = \left\| \frac{dmT^{-1}}{dm} \right\|_\infty^{1/2} ,$$

where $\| \ \|_\infty$ indicates the essential supremum norm.

For the remainder of the proof one observes that absolute continuity of mT^{-1} with respect to m suffices to make C_T a mapping of equivalence classes of functions, and boundedness of dmT^{-1}/dm implies inequalities (5), (4) and (3) with any bound for dmT^{-1}/dm in place of $\|C_T\|^2$. Boundedness of C_T follows as well as the opposite inequality to (5).

Operators C_T, which we will call composition operators, originally appeared in the work of B. O. Koopman [Kp] on classical mechanics, as mentioned previously, (see [KN], [N1], [N2], [HN] and [H1]), and they have played a role in ergodic theory (see [H3]). The initial studies dealt with invertible measure-preserving transformations whereas a number of recent papers have considered more general transformations. It is the later work that we will survey here.

B. Examples. Before taking up questions of characterization and spectral properties we wish to observe that examples of composition operators abound.

1. If T is an invertible measure-preserving transformation of X onto X, then C_T is a unitary operator. For example let X be the real line $\mathbb{R}$ and put $Tx = x-1$, or let X be $[0,1]$, let α be a point of X and put $Tx = x-\alpha$ (mod 1), or let X be the integers $\mathbb{Z}$ and put $Tn = n-1$. The first two of these examples use Lebesgue measure and the third counting measure. These are the types of operators that arose in Koopman's study and are called translation operators. See [H3] for more interesting examples of this type and their properties.

2. In the integer example above $L^2(m)$ is $\ell^2(\mathbb{Z})$ and C_T is the bilateral shift. Taking X to be the nonnegative integers $\mathbb{Z}^+$ and Y the positive integers, we obtain the unilateral shift from the transformation $Tn = n-1$. Replace counting measure on $\mathbb{Z}$ or $\mathbb{Z}^+$ by an arbitrary measure, and $L^2(m)$ becomes a weighted sequence space. The composition operator induced by the same transformation is then a bilateral or unilateral weighted shift. See Shields' excellent article [Sh] for an account of these operators.

3. Suppose X is $\mathbb{R}$ with Lebesgue measure, $a > 0$ and $Tx = ax$. It is then an exercise to work out the properties of C_T. More generally, let T be a monotone function on R. The condition of Theorem 1 is equivalent to $1/T'$ being essentially bounded [Sn5].

4. With X the unit circle in the complex plane $\mathbb{C}$ and m normalized Lebesgue measure on X ($dm = d\theta/2\pi$), let T be a nonconstant inner function, i.e., $|T| = 1$ a.e. and the Poisson integral of T is analytic. We write $T(0)$ for $\int T\,dm$. The Fourier coefficients of mT^{-1} may be evaluated using (2) as follows: for $n \geq 0$,

$$\int e^{in\theta}\, dmT^{-1}(e^{i\theta}) = \int T^n\, dm = \left(\int T\, dm\right)^n = T(0)^n,$$

where the second equality above is a consequence of analyticity (see [Du], Chap. 3). Fourier coefficients of positive index are obtained from the above equality by taking complex conjugates, and thus the Fourier coefficients of mT^{-1} agree with those of the Poisson kernel for evaluation at $T(0)$. Hence $mT^{-1} \ll m$, and

$$\frac{dmT^{-1}}{dm}(e^{i\theta}) = \mathrm{Re}\,\frac{e^{i\theta}+T(0)}{e^{i\theta}-T(0)}. \tag{6}$$

Since

$$\|f\circ T\|^2 = \int |f|^2\,\frac{dmT^{-1}}{dm}\,dm, \tag{7}$$

it follows that C_T is bounded on $L^2(m)$, and

$$\|C_T\|^2 = (1+|T(0)|)/(1-|T(0)|).$$

See [Nr] for further discussion of this example.

These are the principal classes of examples I am familiar with. There must be more that are of interest.

C. Characterization. We turn next to a characterization of composition operators in terms of preservation of pointwise products. The prototype of such a characterization is due to von Neumann [N2] (see also [H3], page 45). The result presented here as Theorem 3 is due to Ridge [R1], [R3]. It depends on the following generalization (Theorem 2) due to Sikorski [Sk1], [Sk2] of von Neumann's characterization [N1] of measure algebra isomorphisms. We need to introduce a little terminology and notation before stating it. (See [H2], Sec. 42 and [Sk2]).

Let $\mathcal{J}$ be a σ-ideal in $\mathcal{S}$. (The only case we are interested in here is the one where $\mathcal{J}$ is the collection of null sets of m.) The σ-algebra $\mathcal{S}$ is a ring in the algebraic sense under the operations of symmetric difference Δ for addition and intersection $\cap$ for multiplication. In fact it is a Boolean σ-algebra, i.e., multiplication is idempotent, $\mathcal{S}$ is closed under the formation of countable unions, and it contains a largest element X. The quotient ring $\mathcal{S}/\mathcal{J}$ is also a Boolean σ-algebra. We denote the equivalence class of a set E in $\mathcal{S}$ by $[E]$.

Let Y be a second set, $\mathcal{T}$ a σ-algebra of subsets of Y and $\mathcal{I}$ a σ-ideal in $\mathcal{T}$. We wish to consider σ-homomorphisms from $\mathcal{S}$ and $\mathcal{S}/\mathcal{J}$ to $\mathcal{T}/\mathcal{I}$, i.e., maps that preserve symmetric differences, intersections, countable unions and maximal elements. A way to obtain such a σ-homomorphism is to take a measurable transformation

$T : Y \to X$ and define $\varphi_T : \mathcal{S} \to \mathcal{T}/\mathcal{J}$ by $\varphi_T(E) = [T^{-1}(E)]$ for E in $\mathcal{S}$. (Here $[T^{-1}(E)]$ denotes the equivalence class of $T^{-1}(E)$ in $\mathcal{T}/\mathcal{J}$.) In case $T^{-1}(E)$ is in $\mathcal{J}$ whenever E is in $\mathcal{I}$ one can define $\Phi_T : \mathcal{S}/\mathcal{I} \to \mathcal{T}/\mathcal{J}$ by $\Phi_T([E]) = \varphi_T(E)$. The σ-homomorphisms φ_T and Φ_T are said to be induced by the transformation T. The point of the following theorem is that this is the only way to obtain a σ-homomorphism under our standing hypothesis that X is a standard Borel space and $\mathcal{S}$ is the algebra of Borel subsets of X.

THEOREM 2. Every σ-homomorphism of $\mathcal{S}/\mathcal{I}$ into $\mathcal{T}/\mathcal{J}$ is induced by a measurable transformation of Y into X.

Sketch of Proof. Let Φ be a σ-homomorphism of $\mathcal{S}/\mathcal{I}$ into $\mathcal{T}/\mathcal{J}$. In case X is countable ($X = \{x_1, x_2, \ldots\}$) it is possible to choose a countable pairwise disjoint collection of sets Y_n whose union is Y such that Y_n is a representative of the image of the singleton $[\{x_n\}]$. It is now possible to define the required T by taking $T(y) = x_n$ for all y in Y_n.

Thus only the case of uncountable X needs to be considered. But there is only one uncountable standard Borel space up to Borel isomorphism (Kuratowski [Kr], p. 451), and thus it suffices to consider the case where X is the Cantor set, which we obtain as the Cartesian product of denumberably many copies of the doubleton $\{0,1\}$. Let Z_n consist of all sequences of 0's and 1's with a 1 as the nth term, and let A_n be a representative of $\Phi([Z_n])$. If χ_n is the characteristic function of A_n, then define T by

$$T(y) = (\chi_1(y), \chi_2(y), \ldots) .$$

Thus $T^{-1}(Z_n) = A_n$, and since the sets Z_n generate $\mathcal{S}$, it follows T is measurable. If $\varphi(E) = \Phi([E])$, then φ is a σ-homomorphism of $\mathcal{S}$ into $\mathcal{T}/\mathcal{J}$ which agrees with φ_T on the sets Z_n, and consequently $\varphi_T = \varphi$. It follows that φ_T sends $\mathcal{I}$ into $\mathcal{J}$ and $\Phi = \Phi_T$. (See [Sk2], pages 139 and 37 for more details.)

THEOREM 3. If m is a σ-finite measure on the Borel subsets of a standard Borel space and A is an operator on $L^2(m)$ such that $A(fg) = AfAg$ whenever f,g and fg are in $L^2(m)$, then A is a composition operator.

<u>Proof</u>. If $m(E) < \infty$, then $\chi_E \in L^2(m)$ and

$$A\chi_E = A\chi_E^2 = (A\chi_E)^2 .$$

Thus there is a set G such that $A\chi_E = \chi_G$. Define φ_0 by $\varphi_0(E) = G$. If $m(F) < \chi$ and $E \cap F = \varphi$, then

$$0 = A\chi_E\chi_F = A\chi_E A\chi_F ,$$

and consequently $m(\varphi_0(E) \cap \varphi_0(F)) = 0$.

Since m is σ-finite, there are pairwise disjoint sets X_n of finite measure such that $X = \bigcup_{n=1}^{\infty} X_n$. Put $Y_n = \varphi_0(X_n)$ and $Y = \bigcup_{n=1}^{\infty} Y_n$. Let

$$\mathcal{J} = \{E \in \mathcal{S} : m(E) = 0\} ,$$

$$\mathcal{T} = \{E \subset Y : E \in \mathcal{S}\} ,$$

and

$$\mathcal{G} = \{E \in \mathcal{T} : m(E) = 0\} .$$

We may suppose $\varphi_0(E) \in \mathcal{T}$ for every E. Thus we obtain a map $\varphi: \mathcal{S} \to \mathcal{T}$ by taking

$$\varphi(E) = \bigcup_{n=1}^{\infty} \varphi(E \cap X_n) .$$

If $E \in \mathcal{J}$, then $\varphi(E) \in \mathcal{G}$, and it therefore makes sense to define $\Phi : \mathcal{S}/\mathcal{J} \to \mathcal{T}/\mathcal{G}$ by

$$\Phi([E]) = [\varphi(E)] .$$

Note that in case m is finite and $A\chi_E = \chi_F$, Φ is simply the function that assigns $[F]$ to $[E]$.

It is routine but tedious to verify that Φ is a σ-homomorphism. By Theorem 2, there is a measurable transformation T from Y to X that induces Φ. If $m(E) < \infty$, then it follows that

$$A\chi_E = \chi_{T^{-1}_E} .$$

Consequently, T induces a composition operator, and $A = C_T$.

Note the multiplicativity hypothesis was only applied to characteristic functions.

On the basis of the theorem we can observe that C_T is invertible only if the transformation T is. To say T is invertible means there exists a measurable transformation U such that $U \circ T$ and $T \circ U$ differ from the identity at most on a set of measure zero.

COROLLARY. (Singh [Sn4].) If C_T is an invertible operator, then T is invertible and C_T^{-1} is induced by any inverse of T.

Proof. Suppose $C_T f = \chi_E$, and $F = \{x : f(x) \neq 0,1\}$. Then $T^{-1}F = \{y : f(Ty) \neq 0,1\}$, which is a null set, and $C_T \chi_F = \chi_{T^{-1}F} = 0$. Consequently $\chi_F = 0$, and f is a characteristic function. Thus C_T^{-1} preserves the class of characteristic functions, and that it preserves products in this class follows from the corresponding property for C_T. By the remark following the proof of the theorem, C_T^{-1} is a composition operator, say C_U.

We must show

$$U \circ T(x) = T \circ U(x) = x \quad \text{a.e.}$$

Since $C_{U\circ T} = C_T C_U = 1$, it is clear that $f \circ U \circ T = f$ for all f in $L^2(m)$. If X is countable, then the desired conclusion is immediate. In the uncountable case we need only, as previously, consider the case when X is the Cantor set. Write X as a disjoint union of sets X_n of finite measure, and let Z_n consist of all sequences of 0's and 1's with n^{th} term 1, as before. The characteristic function χ_{mn} of $Z_m \cap X_n$ is in $L^2(m)$, and thus $\chi_{mn} \circ U \circ T(x) = \chi_{mn}(x)$ a.e. Since the countable set of functions $\{\chi_{mn}\}$ separates points of X, the conclusion follows in this case also.

The converse to the corollary is false. Take $T(x) = \sqrt{x}$ on $L^2(0,1)$. Then as in Example 3, C_T is an operator. But $T^{-1}(x) = x^2$, and T^{-1} does not induce an operator, since the pertinent Radon-Nikodym derivative is the unbounded function $1/(2\sqrt{x})$. Singh [Sn4] has given an example of an invertible T such that C_T has nontrivial kernel.

Singh has also observed [Sn2] that if $\varphi = dmT^{-1}/dm$, then $C_T^* C_T = M_\varphi$, and consequently the kernel of C_T consists precisely of the functions supported on the set where φ vanishes.

D. Spectra. My final remarks in this part concern symmetry of spectra of composition operators. Weighted shifts are prime examples of composition operators, so one might hope for their spectral properties to carry over (see [Kl], [Sh]). That circular symmetry is too much to expect is demonstrated by the example $\begin{bmatrix} 0 & 1 & 0 \\ 0 & 0 & 1 \\ 1 & 0 & 0 \end{bmatrix}$, which is a composition operator with a three point spectrum. In the positive direction we have the following theorem due to Ridge [R2]. We use Λ, Π and Π_0 to denote spectrum, approximate point spectrum and point spectrum, respectively.

THEOREM 4. (i) If $|\lambda| = 1$ and $\lambda \in \Pi_0(C_T)$, then $\lambda^n \in \Pi_0(C_T)$ for every integer n. The same is true of Π.

(ii) If $|\lambda| \neq 1$ and $\lambda \in \Pi_0(C_T)$, then $e^{i\theta}\lambda \in \Pi_0(C_T)$ for every real θ. The same is true of Π and Λ.

(iii) If $m(X) < \infty$, $\lambda > 0$, $\lambda \neq 1$ and $\lambda \in \Pi_0(C_T)$, then $\Pi_0(C_T)$ includes the open annulus centered at zero determined by 1 and λ.

Proof. Only the proof of the first statement in (ii) is included to give an idea of the construction. The point spectrum proofs are all somewhat similar, and the approximate point spectrum proofs are "epsilonic" modifications of these. See Ridge's paper [R2] for details.

Suppose $\lambda \in \Pi_0(C_T)$, $|\lambda| \neq 0,1$. Let f be a corresponding eigenvector: $f \circ T = \lambda f$. On taking absolute values, we see

$$|f| \circ T = r|f| .$$

Given real θ, put

$$g = |f| \exp(i\,\theta \log|f|/\log r) .$$

An easy calculation shows $g \circ T = r\, e^{i\theta} g$.

Part II. Functional Hilbert Spaces, H^2

A. Functional Hilbert Spaces. We will first characterize composition operators on functional Hilbert spaces and then specialize to H^2 of the disc. A Hilbert space of complex valued functions on a set X is called a functional Hilbert space if the operations of addition and scalar multiplication are the pointwise ones and if each point evaluation $(f \to f(x))$ is a bounded linear functional (see [H4]). If H is a functional Hilbert space and if T maps X into X in such a manner that $f \circ T \in H$ whenever $f \in H$, then it follows easily from the closed graph theorem that the linear transformation C_T sending f to $f \circ T$ is in fact a bounded operator. We call C_T a composition operator on the functional Hilbert space H.

By the Riesz representation theorem, corresponding to each point x of X there is a vector k_x such that for f in H,

$$f(x) = (f, k_x) .$$

Multiplication operators on H are characterized by their adjoints having the k_x's as eigenvectors (cf. [SW], p. 783). For composition operators Caughran and Schwartz [CS] observed the following:

THEOREM 1. An operator A on H is a composition operator if and only if the set $\{k_x : x \in X\}$ is invariant under A^*. In this case T is determined by $A^*k_x = k_{Tx}$.

Proof. If $A = C_T$, then for every f in H,

$$(f, A^*k_x) = (Af, k_x) = f(T(x)) = (f, k_{Tx}) .$$

Consequently $A^*k_x = k_{Tx}$.

Conversely, if $A^*k_x = k_{Tx}$, then

$$Af(x) = (Af, k_x) = (f, A^*k_x) = f(Tx) .$$

Hence $A = C_T$.

Maps that induce composition operators may be characterized in terms of the kernel function K of H, which is defined by

$$K(x,y) = (k_y, k_x) .$$

Let $K^{\sim}(x_1,\ldots,x_n)$ be the matrix with entries $K(x_i,x_j)$. Being a Gramm matrix, $K^{\sim}$ is positive.

THEOREM 2. A map T of X into itself induces a composition operator on H if and only if there is a constant M such that

$$(1) \qquad K^{\sim}(Tx_1,\ldots,Tx_n) \leq M\, K^{\sim}(x_1,\ldots,x_n)$$

for every finite subset $\{x_1,\ldots,x_n\}$ of X. In this case $\|C_T\|^2$ is the best possible constant.

<u>Proof</u>. Let $\{x_1,\ldots,x_n\}$ be a subset of X and let $\{\alpha_1,\ldots,\alpha_n\}$ be a corresponding set of complex numbers. We will write k_i for k_x when $x = x_i$ and k_i' for k_x when $x = Tx_i$.

If C_T is an operator, then by the preceding theorem $C_T^* k_i = k_i'$, and consequently

$$(2) \qquad \left\| \sum_{i=1}^{n} \alpha_i k_i' \right\|^2 \leq M \left\| \sum_{i=1}^{n} \alpha_i k_i \right\|^2$$

with $M = \|C_T^*\|^2 = \|C_T\|^2$. Since the set of finite linear combinations of the form $\sum_{i=1}^{n} \alpha_i k_i$ is dense in H, it follows that T induces a bounded operator only if (2) holds for all choices of x_i and α_i, and in this case $\|C_T\|^2 \leq M$. But (2) is equivalent to

$$\sum_{i,j=1}^{n} \bar{\alpha}_i \alpha_j\, K(Tx_i,Tx_j) \leq M \sum_{i,j=1}^{n} \bar{\alpha}_i \alpha_j\, K(x_i,x_j) ,$$

which is in turn equivalent to (1).

For later use we record a lemma here.

LEMMA . A sequence in a functional Hilbert space is a weak null sequence if and only if it is norm bounded and a pointwise null sequence.

Proof. Necessity follows from the principle of uniform boundedness and the fact that point evaluations are inner products. Suppose conversely that $\{f_n\}$ is a sequence in H, $\|f_n\| \le M$ for all n and $\lim_{n\to\infty} f_n(x) = 0$ for every x. Since the functions k_x form a spanning set, an arbitrary vector g can be approximated to within an arbitrary $\varepsilon > 0$ by a finite linear combination of k_x's:

$$g = g_0 + \sum_{x\in F} \alpha_x k_x ,$$

where $\|g_0\| < \varepsilon$ and F is finite. It follows that

$$|(f_n,g)| \le |(f_n,g_0)| + |(f_n, \sum_{x\in F} \alpha_x k_x)|$$

$$< M\varepsilon + \sum_{x\in F} |\alpha_x| \, |f_n(x)| ,$$

and hence $\lim_{n\to\infty} (f_n,g) = 0$.

B. $H^2(D)$. We turn to the special case in which $H = H^2(D)$, where D is the open unit disc in the complex plane. There are several equivalent ways to define $H^2(D)$. The simplest is that $f \in H^2(D)$ if and only if

$$f(z) = \sum_{n=0}^{\infty} a_n z^n \quad \text{for } z \text{ in } D$$

and $\sum_{n=0}^{\infty} |a_n|^2 < \infty$. A method of Rudin [Rd1], [Rd2] better adapted to our needs involves harmonic majorants. An analytic function f on D is in $H^2(D)$ if and only if there is a harmonic function h, called a harmonic majorant for f, such that

$$|f(z)|^2 \le h(z) \quad \text{for } z \text{ in } D .$$

If f has a harmonic majorant h, then it has a smallest one h_f, called the least harmonic majorant, i.e., h_f is a harmonic majorant for f and $h_f \le h$ for every other harmonic majorant h. We have $\|f\|^2 = h_f(0) = \sum_{n=0}^{\infty} |a_n|^2$. See [Rd2] and [Du] for the details.

Let T be an analytic function on D whose range is in D. If $f \in H^2(D)$, then $f \circ T$ is analytic, $h_f \circ T$ is harmonic, and clearly

$$|f(T(z))|^2 \le h_f(T(z)) \quad \text{for } z \text{ in } D,$$

i.e., $f \circ T$ has a harmonic majorant. Thus for each f in H^2, $f \circ T \in H^2(D)$, and consequently C_T is an operator on $H^2(D)$. Further, since $h_f \circ T$ is a harmonic majorant for f,

$$h_{f \circ T} \le h_f \circ T,$$

and thus

$$\|f \circ T\|^2 = h_{f \circ T}(0) \le h_f(T(0)).$$

Applying Harnack's inequality to $h_f(T(0))$, we see

$$\|f \circ T\|^2 \le \frac{1 + |T(0)|}{1 - |T(0)|} h_f(0) = \frac{1 + |T(0)|}{1 - |T(0)|} \|f\|^2,$$

and consequently

$$\|C_T\|^2 \le \frac{1 + |T(0)|}{1 - |T(0)|}.$$

We may summarize the foregoing as follows.

THEOREM 3. If T is an analytic function mapping the unit disc into itself, then T induces a composition operator on $H^2(D)$ of norm no greater than $[(1 + |T(0)|)/(1 - |T(0)|)]^{1/2}$.

This result goes back to Littlewood [L] in 1925 when $T(0) = 0$. It was obtained in its present form by Ryff [Rf] in 1966, but by techniques different from ours. The technique of employing harmonic majorants to obtain boundedness has also been used by Swanton [Sw] in studying composition operators on multiply connected domains.

A special case of Theorem 3 was observed for inner functions in Part I, Example 4. In that case the inequality just derived turned out to be equality. That this is not always the case here may be seen by examining the operator induced by the constant function with value 1/2 for example.

A well known theorem of Fatou asserts the existence of radial limits for functions f in $H^2(D)$: for almost every θ, the limit $\lim_{r \to 1-} f(re^{i\theta})$ exists and defines a function $f^\wedge(e^{i\theta})$ on the unit circle ∂D. Thus f can be extended to the closed disc except possibly for a set of measure zero on the boundary. I will denote this extension by f also, relying on context to distinguish the original function from its extension. The map sending f to $f^\wedge$ is an isometry of $H^2(D)$ onto a subspace $H^2(\partial D)$ of $L^2(m)$, where m is normalized Lebesgue measure on the circle. As is well known, f may be recaptured from $f^\wedge$ by means of the Poisson integral. Again see [Du] for details.

In particular if T is analytic and maps D into D, then it has a boundary function $T^\wedge$, and we may ask about the relationship between $(f \circ T)^\wedge$ and $f \circ (T^\wedge)$. (Note that in the first expression f is a function on D whereas in the second f is the extension to $\bar{D}$.) Ryff proved all is well.

THEOREM 4. $(f \circ T)^\wedge (e^{it}) = f(T^\wedge(e^{it}))$ a.e.

Proof. See [Rf].

C. Characterization and Invertibility. Composition operators on $H^2(D)$ may be characterized in essentially the same way as those on $L^2(m)$, Part I, Theorem 3. Recall that the functions e_n defined by $e_n(z) = z^n$, $n = 0,1,\ldots$, form a basis for $H^2(D)$.

THEOREM 5. A nonzero operator A on $H^2(D)$ is a composition operator if and only if $Ae_n = (Ae_1)^n$ for $n = 0,1,2,\ldots$.

Proof. Necessity is trivial. As for sufficiency, let $T = Ae_1$. Then

$$\|T^n\| = \|Ae_n\| \leq \|A\| .$$

Taking n^{th} roots and limits as $n \to \infty$, we see that the right hand side tends to 1, and if the H^2 norms are written as integrals on the circle, the left tends to $\|T^\wedge\|_\infty$. Thus by the maximum modulus theorem, either T maps D into D, or T is constant with unit modulus. The latter is impossible, since the bounded operator A cannot send every basis vector e_n to $\alpha^n e_0$ with $|\alpha| = 1$. Thus T induces a composition operator which agrees with A on a basis. Hence $A = C_T$.

COROLLARY. If A is a nonzero operator on $H^2(D)$ such that $Afg = Af\,Ag$ whenever f and g are bounded, then A is a composition operator.

The above results and the following one are due to Schwartz [Sc]. Peter Rosenthal has observed that essentially the same proof used above serves to establish Theorem 3 of Part I in the special case where m is Lebesgue measure on the circle.

THEOREM 6. A composition operator C_T on $H^2(D)$ is invertible if and only if T is a Möbius transformation of D onto D.

<u>Proof</u>. Sufficiency is trivial, so suppose C_T is invertible with inverse A. Let f and g be bounded functions in $H^2(D)$. Then

$$(Afg) \circ T = fg = (Af \circ T)(Ag \circ T) = (AfAg) \circ T ,$$

or equivalently

$$(Afg - AfAg) \circ T = 0 .$$

Invertibility of C_T implies T is nonconstant, and thus the range of T is an open subset of D. Hence the last equation implies

$$Afg = AfAg .$$

By the corollary, $A = C_U$ for some U. Applying $C_U C_T$ and $C_T C_U$ to z, we see that T is an invertible analytic map of D onto D, and hence it is a Möbius transformation.

The literature contains a number of other results concerning invertibility, which we mention here without proof. The simplest observation along these lines is that C_T has a trivial kernel unless T is constant. A nontrivial result of Cima, Thomson and Wogen [CTW] shows that C_T is seldom Fredholm.

THEOREM 7. If C_T is Fredholm, then it is invertible.

In the same paper they obtain a necessary and sufficient condition for the range of C_T to be closed. It is that $d(m\hat{T}^{-1})/dm$ be bounded away from zero, which roughly says that the range of $T^{\wedge}$ includes the unit circle in a strong way. They give an example to show that inclusion alone does not suffice. Thus there are many semi-Fredholm composition operators, but they are either invertible or of index $-\infty$.

Conditions under which the range of C_T is dense have been obtained by J. Caughran and R. C. Roan. It is easy to see that C_T has dense range only if T is one-to-one. For if $T(z) = T(w)$, then $C_T^*(k_z - k_w) = 0$ (Theorem 1), and hence $k_z - k_w$ is orthogonal to the range of C_T. Thus we restrict our attention to one-to-one T, but we will see that this condition alone does not suffice for the range of T to be dense. The following is due to Caughran [Ca].

Let G be the range of T, so G is a simply connected open subset of D. We can define $H^2(G)$ by the same method used to define $H^2(D)$. Thus an analytic function f on G is in $H^2(G)$ if and only if $|f|^2$ is dominated by a harmonic function on G. The least harmonic majorant of f determines a norm for f given by

$$\|f\| = h_f(T(0))^{1/2} .$$

With this definition the mapping that sends f in $H^2(G)$ to $f \circ T$ in $H^2(D)$ is clearly invertible and norm-preserving. Thus the range of C_T is dense in $H^2(D)$ if and only if the polynomials $p(T)$ in T are dense in $H^2(D)$ if and only if the polynomials p are dense in $H^2(G)$.

Density of the polynomials in $H^2(G)$ can be related to geometrical properties of G. We say G is a Caratheodory domain provided $\partial G = \partial(C/\overline{G})_\infty$, where A_∞ is the unbounded component of A, and we say G is a Jordan domain if it is the interior of simple closed curve.

THEOREM 8. If G is a Caratheodory domain, then the polynomials are dense in $H^2(G)$. If T is continuous on $\overline{D}$ and the polynomials are dense in $H^2(G)$, then G is a Jordan domain.

THEOREM 9. If T is one-to-one and $T(D)$ is a Caratheodory domain, then the range of C_T is dense. If T is continuous on $\overline{D}$ and the range of C_T is dense, then $T(D)$ is a Jordan domain.

On the basis of Theorem 9, an example may be constructed of a one-to-one T such that the range of T is not dense. Let G be the disc D minus the unit interval, and let T map D conformally onto G. Then T can be extended continuously to $\overline{D}$, but the range of C_T cannot be dense since G is not a Jordan domain.

An extension of Theorem 9 has recently been obtained by R. C. Roan [Rn3]. A function φ is a weak* generator of H^{∞} if the polynomials in φ are weak* dense in H^{∞} (see [Sa]).

THEOREM 10. If T is a weak* generator of H^{∞}, then C_T has dense range.

D. Compactness. Compactness questions have inspired some of the deepest studies of composition operators. In this section we will sample some of the results produced in attempting to characterize the compact composition operators on $H^2(D)$. The problem still lacks a complete solution. We will also see that compactness of a composition operator implies the existence of a fixed point for the function that induces it.

The first result in this area is due to Schwartz [Sc].

THEOREM 11. If C_T is compact, then $|T^{\wedge}| < 1$ a.e.

Proof. Since the range of T is in D, it is clear that $|T^{\wedge}| \leq 1$ a.e. Because $\{e_n : n = 0,1,2,\ldots\}$ is orthonormal and C_T is compact, we have $\|C_T e_n\| \to 0$ as $n \to \infty$. But

$$\|C_T e_n\|^2 = \int |T^{\wedge}|^{2n}\, dm ,$$

so it follows that $T^{\wedge}$ cannot have modulus one on a set of positive measure. Thus $|T^{\wedge}| < 1$ a.e.

Schwartz gave an example showing the above condition is not sufficient. We will look at his example after Theorem 13. He also gave a sufficient condition for compactness which Shapiro and Taylor [ST] noted is characteristic of Hilbert-Schmidt composition operators.

THEOREM 12. A composition operator is Hilbert-Schmidt if and only if $1/(1-|T^{\wedge}|) \subset L^1(m)$.

Proof. Since $1 \leq 1 + |T^{\wedge}| \leq 2$, $1/(1-|T^{\wedge}|) \in L^1(m)$ if and only if $1/(1-|T^{\wedge}|^2) \in L^1(m)$. The operator C_T is Hilbert-Schmidt if and only if $\sum_{n=0}^{\infty} \|C_T e_n\|^2 < \infty$, and hence the result is a consequence of the following calculation:

$$\sum_{n=0}^{\infty} \|C_T e_n\|^2 = \sum_{n=0}^{\infty} \int |T_{\infty}|^{2n} \, dm = \int 1/(1-|T^{\wedge}|^2) \, dm .$$

COROLLARY 1. If $r < 1$ and $|T^{\wedge}| \leq r$ a.e., then C_T is trace class.

That C_T is compact under the hypothesis of the corollary was observed by Schwartz. The trace class conclusion was obtained by Shapiro and Taylor [ST] on the basis of the following very useful observation.

PROPOSITION. Suppose T is one-to-one and C_T belongs to some left ideal of operators. If U is an analytic function on D whose range is included in that of T, then C_U belongs to the same left ideal.

Proof. Define V on D by $V = T^{-1} \circ U$, so $U = T \circ V$. The range of V is also in D, and hence V induces a composition operator satisfying

$$C_U = C_{T \circ V} = C_V C_T .$$

The result follows.

Proof of Corollary 1. Define R on D by $R(z) = \sqrt{r}\, z$. That C_R is Hilbert-Schmidt follows from the theorem, and thus $C_{R \circ R} (= C_R^2)$ is trace class. By hypothesis, the range of T is included in the range of the one-to-one function $R \circ R$. Hence the corollary follows from the proposition.

COROLLARY 2. [ST] If the range of T is included in a polygon inscribed in the unit circle, then C_T is a trace class operator.

Proof. We will only sketch that part of Shapiro's and Taylor's proof that shows C_T is Hilbert-Schmidt and refer the interested reader to their paper for the full result. By the proposition, it will suffice to show that $1/(1-|T^\wedge|) \in L^1(m)$ in the case where T is a conformal mapping of D onto the given polygon. Further, it is enough to demonstrate integrability in some neighborhood of the preimage of an arbitrary vertex. With no loss of generality we suppose the vertex is at 1 and $T(1) = 1$. For suitable $\gamma > 1$, depending on the angle between the two sides of the polygon that meet at 1, $(1-T(z))^\gamma$ maps 1 to 0 and a neighborhood of 1 in D onto the intersection of a half plane with a neighborhood of 0. By the Schwarz reflection principal, $(1-T(z))^\gamma$ has an analytic extension to a neighborhood of 1, and

$$(1-T(z))^\gamma = (1-z)h(z) ,$$

where $h(z)$ is analytic at 1 and $h(1) \neq 0$. If $\alpha = 1/\gamma$, then

$$1-T(z) = (1-z)^\alpha h(z)^\alpha ,$$

and on a suitable small neighborhood of 1 with suitable constants r and s ($s \leq |h(z)|$ in the neighborhood), we have

$$1-|T^\wedge(e^{i\theta})| \geq r|1-T^\wedge(e^{i\theta})| \geq rs|1-e^{i\theta}|^\alpha .$$

Thus $1/|1-T^\wedge|$ is integrable near 1.

Another sufficient condition for C_T to be Hilbert-Schmidt obtained by Shapiro and Taylor is that $(1-|z|)^{-3}$ be integrable over the range of T with respect to area measure. Thus one can give examples of functions T inducing Hilbert-Schmidt composition operators such that the range of T has infinitely many spikes extending out to the unit circle or such that the range of T is an infinite ribbon spiraling out to the unit circle. Thus although Theorem 11 implies that for C_T to be compact $T(z)$ must stay away from the unit circle in some sense, these examples show that the relationship is delicate. The next theorem gives further indication of how delicate this relationship is, but we need to give a definition before it can be stated.

A function T on D is said to have an angular derivative at $e^{i\theta}$ if there exist complex numbers c and $e^{i\omega}$ such that $(\varphi(z)-e^{i\omega})/(z-e^{i\theta})$ tends to c

as z tends to $e^{i\theta}$ over any triangle with interior in D and one vertex at $e^{i\theta}$. The following result and the proof given are also due to Shapiro and Taylor [ST].

THEOREM 13. If T has an angular derivative at one point of the unit circle, then C_T is not compact.

Proof. With no loss of generality we may assume $e^{i\theta} = e^{i\omega} = 1$, and thus there is a constant M such that

(3) $$|1 - T(x)|/|1 - x| \leq M \quad \text{for} \quad -1 < x < 1 .$$

Let

$$f_n(z) = n^{-1/2}(1 - z)^{-(1-1/n)/2}$$

for z in D. Since $(e^{i\theta} - 1)/\theta$ is continuous on $[-\pi,\pi]$ and never vanishes, it is bounded away from 0, and thus there is a constant k such that

$$|1 - e^{i\theta}|^{-1} \leq k|\theta|^{-1}$$

for $0 < |\theta| < \pi$. Hence

$$\|f_n\|^2 = (2\pi)^{-1} \int_{-\pi}^{\pi} n^{-1}|1 - e^{i\theta}|^{(1/n)-1}\, d\theta$$

$$\leq (2\pi n)^{-1} \int_{-\pi}^{\pi} (k|\theta|)^{(1/n)-1}\, d\theta = (k\pi)^{(1/n)-1} .$$

Thus (f_n) is a bounded sequence in $H^2(D)$. It is clearly a pointwise null sequence, and thus the lemma of Section A implies it is a weak null sequence.

The proof is completed by showing $(C_T f_n)$ is bounded away from zero. For this we apply the Fejer-Riesz inequality to $C_T f_n$, obtaining

(4) $$\|C_T f_n\|^2 \geq \frac{1}{\pi} \int_{-1}^{1} |f_n \circ T(x)|^2\, dx .$$

But by using the inequality (3), we see

$$|f_n \circ T(x)|^2 = n^{-1}\, |1 - T(x)|^{-(1-1/n)} \geq n^{-1}\, [M(1 - x)]^{-(1-1/n)} ,$$

and hence the integral in (4) is at least as large as $\pi^{-1}M^{1/n-1}\,2^{1/n}$, which establishes that $\{C_T f_n\}$ is bounded away from zero, and hence C_T is not compact.

In particular the theorem implies that if $T(1) = 1$ and T is analytic at 1, then C_T is not compact. Schwartz's example, referred to earlier, is the function $T(z) = (1 + z)/2$. The image of T touches the circle at just one point, but it is immediate from Theorem 13 that C_T is not compact. It is an open question whether the converse of Theorem 13 is true. Swanton showed in his thesis [Sw] that the converse is true if compactness of C_T is implied by total boundedness of the set $\{k_{Tz}/\|k_z\| : z \in D\}$, where k_z is the kernel function for evaluation at z. We conclude this section with a result of Caughran and Schwartz [CS].

THEOREM 14. If C_T^N is compact for some N, then T has a fixed point in D.

Proof. We consider first the case when C_T is itself compact. By Theorem 6, T is not a Möbius transformation. A theorem of Wolff [W] and Denjoy [Dn] asserts that if an analytic function T mapping D into D is not just a rotation about a fixed point, then there is a constant a in $\overline{D}$ to which the iterates of T converge (uniformly on compact subsets of D). Writing T^n for the composite of T with itself n times, we see there is a point a in $\overline{D}$ such that $T^n(0) \to a$ as $n \to \infty$. If $a \in D$, then a is the required fixed point.

It will be shown that a contradiction results from supposing $|a| = 1$. A calculation shows that the kernel function for $H^2(D)$ is given by

$$k_\zeta(z) = 1/(1 - \overline{\zeta}\, z)\,,$$

and hence $\|k_\zeta\|^2 = 1/(1 - |\zeta|^2)$ (see [H4], Problem 30). Let $\zeta_n = T^n(0)$, and write k_n for k_ζ when $\zeta = \zeta_n$. Our supposition implies $\lim_{n\to\infty} |\zeta_n| = 1$. Hence, for z in D,

$$\lim_{n\to\infty} k_n(z)/\|k_n\| = 0\,,$$

and thus the lemma of Section A implies $\{k_n/\|k_n\|\}$ is a weak null sequence. Since

C_T^* is compact, and, by Theorem 1, $C_T^* k_n = k_{n+1}$, it follows that $\{k_{n+1}/\|k_n\|\}$ is a strong null sequence, i.e.,

$$\lim_{n\to\infty} (1 - |\zeta_n|^2)/(1 - |\zeta_{n+1}|^2) = 0 .$$

But $|\zeta_n| \to 1$ implies that $|\zeta_{n+1}| \geq |\zeta_n|$ infinitely often, and hence we obtain the contradiction

$$\limsup_{n\to\infty} (1 - |\zeta_n|^2)/(1 - |\zeta_{n+1}|^2) \geq 1 .$$

Finally, if $N > 1$, then the above argument may be applied to obtain a fixed point a of T^N in D. As above $\{T^n(a)\}$ is convergent, but in addition $T^{nN}(a) = a$ for all n. Thus $T(a) = a$.

E. Spectra. In this section we will describe what is known about spectra of composition operators on $H^2(D)$. Most proofs will be omitted. We consider five cases.

1. [Rf], [Nr] If T is inner and has a fixed point in D, then C_T is similar to an isometry. For we can define a Möbius transformation β of D onto D taking the fixed point to 0, and thus $U = \beta \circ T \circ \beta^{-1}$ has a fixed point at 0. By equations (6) and (7) of Section B in Part I, C_U is an isometry and $C_U = C_\beta^{-1} C_T C_\beta$. If T is not itself a Möbius transformation of D onto D, then C_T and C_U are not invertible. It follows that $\Lambda(C_T)$ is the closed unit disc.

2. [Nr] If T is a Möbius transformation of D onto D, then the behavior of T depends on its fixed points (see [F] for example). There can be one fixed point inside the unit circle and one outside, in which case T is a generalized rotation and is called elliptic. The U constructed in the preceding case then has the form $U(z) = e^{i\theta}z$ for some φ and $\Lambda(C_T)$ is the closure of $\{e^{in\theta} : n = 0,1,\ldots\}$. Next there can be two fixed points on the unit circle, in which case T is a generalized homothety and is called hyperbolic. The derivative of a hyperbolic T has a positive value $K > 1$ at one fixed point and $\Lambda(C_T)$ is the annulus centered at 0 with outer radius $K^{1/2}$ and inner radius $K^{-1/2}$. There is only one other case, where T has one fixed point on the unit circle,

and in this case T is a generalized translation and is called parabolic. The spectrum of C_T in this case is the unit circle.

3. [CS] Suppose some power of C_T is compact and 0 is the fixed point of T (see Theorem 14). The spectrum of C_T is the closure of the set consisting of $0,1$ and the powers of $T'(a)$.

We sketch the proof. By taking a similarity as in the first case we may assume $a = 0$. The matrix of C_T^* with respect to the basis $\{e_n : n = 0,1,2,\dots\}$ is easily seen to be upper triangular with 1 and the powers of $\overline{T'(0)}$ on the main diagonal, and thus the set described is included in the spectrum. Since some power of T is compact, nonzero points in the spectrum of C_T are eigenvalues. Thus it remains to show that if f is a nonconstant analytic function and $f \circ T = \lambda f$ for $\lambda \neq 1$, then λ is a power of $T'(0)$. This was originally established by Koenig [Kn]. Evaluation of both sides of the last equation at 0 shows $f(0) = 0$. If $T = e_1\omega$ and $f = e_n h$ where $h(0) \neq 0$, then a calculation shows that

$$e_n \omega^n \, h \circ T = \lambda \, e_n \, h .$$

The result follows on cancelling e_n and evaluating at 0.

The remaining two known cases are due to Kamowitz [Km2]. His methods are deeper than any used heretofore in this section, and we will content ourselves with stating his results. The principal restriction is that T must be assumed analytic on $\overline{D}$. We will further assume that no power of C_T is compact and that T is not a Möbius transformation of D onto D since these cases are covered by 2 and 3 above.

4. Suppose T has no fixed point in D. By the Brouwer fixed point theorem, T has one or more fixed points on ∂D. It turns out there is a unique one, say z_0, for which $T'(z_0) \in (0,1]$. Let $b = T'(z_0)$. If $b < 1$, then $\Lambda(C_T)$ is the zero-centered disc of radius $1/b^{1/2}$. If $b = 1$, then it is only known that $\Lambda(C_T)$ is included in the unit disc. This result includes the special case $T(z) = bz + a$, with $b > 0$ and $a + b = 1$, which was originally obtained by

Deddens [D]. He also obtained the spectrum of C_T under the less restrictive requirement $|a| + |b| \leq 1$. This situation turns out to be covered either by the special case or by 3 above.

5. Suppose T has a fixed point z_0 in D and T is not inner. Then $F = \bigcap_{n=0}^{\infty} T^n \partial D$ is a finite subset of ∂D and T permutes the points of F among themselves. Let N be the order of the permutation, and let $c = \min\{(T^N)'(z) : z \in F\}$. Then $c > 1$ and $\Lambda(C_T)$ consists of the zero centered disc of radius $1/c^{1/2N}$ together with the points $1, T'(z_0), T'(z_0)^2, \ldots$.

This concludes the summary of known cases. The spectrum of C_T remains to be investigated when T fails to be analytic on ∂D and T is not an inner function with a fixed point. Also whether the spectrum of C_T can be a proper subset of $\overline{D}$ in Case 4 when $b = 1$ remains to be determined.

REFERENCES

[AB] M. B. Abrahamse and J. A. Ball, Analytic Toeplitz operators with automorphic symbol II, to appear.

[B] J. A. Ball, Hardy space expectation operators and reducing subspaces, Proc. Amer. Math. Soc. 47 (1975), 351-357. MR 50 #10887.

[Ba] J. R. Baxter, A class of ergodic transformations having simple spectrum, Proc. Amer. Math. Soc. 27 (1971), 275-279. MR 43 #2187.

[Be] J. M. Belley, Spectral properties for invertible measuring preserving transformations, Canad. J. Math. 25 (1973), 806-811.

[Bo1] D. M. Boyd, Composition operators on the Bergman space and analytic function spaces on the annulus, Thesis, U. North Carolina, 1974.

[Bo2] D. M. Boyd, Composition operators on the Bergman space, Coll. Math. 34 (1975), 127-136. MR 53 #11416.

[Bo3] D. M. Boyd, Composition operators on $H^p(A)$, Pacific J. Math. 62 (1976), 55-60.

[Ca] J. G. Caughran, Polynomial approximation and spectral properties of composition operators on H^2, Indiana U. Math. J. 21 (1971), 81-84. MR 44 #4213.

[CS] J. G. Caughran and H. J. Schwartz, Spectra of compact composition operators, Proc. Amer. Math. Soc. 51 (1975), 127-130. MR 51 #13750.

[Ch1] J. R. Choksi, Nonergodic transformations with discrete spectrum, Illinois J. Math. 9 (1965), 307-320. MR 30 #4093.

[Ch2] J. R. Choksi, Unitary operators induced by measure preserving transformations, J. Math. Mech. 16 (1966), 83-100. MR 34 #1844.

[Ch3] J. R. Choksi, Unitary operators induced by measurable transformations, J. Math. Mech. 17 (1967/68), 785-801. MR 36 #2003.

[CTW] J. A. Cima, J. Thomson and W. Wogen, On some properties of composition operators, Indiana U. Math. J. 24 (1974), 215-220. MR 50 #2979.

[CW] J. A. Cima and W. Wogen, On algebras generated by composition operators, Canad. J. Math. 26 (1974), 1234-1241. MR 50 #2978.

[D] J. A. Deddens, Analytic Toeplitz and composition operators, Canad. J. Math. 24 (1972), 859-865. MR 46 #9789.

[Dn] M. A. Denjoy, Sur l'iteration des fonctions analytiques, C. R. Acad. Sci. Paris 182 (1926), 255-257.

[Du] P. L. Duren, Theory of H^p Spaces, Academic Press, New York, 1970. MR 42 #3552.

[F] L. Ford, Automorphic Functions, 2nd ed., Chelsea, N.Y. 1951.

[H1] P. R. Halmos, Measurable transformations, Bull. Amer. Math. Soc. 55 (1949), 1015-1034. MR 11-373.

[H2] P. R. Halmos, Measure Theory, Van Nostrand, Princeton, N.J., 1950. MR 11-504.

[H3] P. R. Halmos, Lectures on Ergodic Theory, Chelsea, N.Y., 1956. MR 20 #3958.

[H4] P. R. Halmos, A Hilbert Space Problem Book, Van Nostrand, Princeton, N.J., 1967. MR 34 #8178.

[HN] P. R. Halmos and J. von Neumann, Operator methods in classical mechanics, Ann. Math. 43 (1942), 332-350. MR 4-14.

[Ha] F. Hartman, Inclusion theorems for Sonnenschein matrices, Proc. Amer. Math. Soc. 21 (1969), 513-519. MR 39 #5984.

[Iw] A. Iwanik, Pointwise induced operators on L_p-spaces, Proc. Amer. Math. Soc. 58 (1976), 173-178.

[Km1] H. Kamowitz, The spectra of endomorphisms of the disc algebra, Pacific J. Math. 46 (1973), 433-440. MR 49 #5918.

[Km2] H. Kamowitz, The spectra of composition operators on H^p, J. Functional Anal. 18 (1975), 132-150. MR 53 #11417.

[KS] H. Kamowitz and S. Scheinberg, The spectrum of automorphisms of Banach algebras, J. Functional Anal. 4 (1969), 268-276. MR 40 #3316.

[KM] S. Karlin and J. McGregor, Spectral theory of branching processes, Z. Wahrscheinlichkeitstheorie 5 (1966), 6-33. MR 34 #5167.

[Kl] R. L. Kelley, Weighted shifts on Hilbert space, Thesis, U. Michigan, 1966.

[Kn] M. Koenig, Recherches sur les integrals de certains equations fonctionelles, Annales de l'Ecole Normale 3 (1884), 1-112.

[Kp] B. O. Koopman, Hamiltonian systems and transformations in Hilbert space, Proc. Nat. Acad. Sci. U.S.A. 17 (1931), 315-318.

[KN] B. O. Koopman and J. von Neumann, Dynamical systems of continuous spectra, Proc. Nat. Acad. Sci. U.S.A. 18 (1932), 255-263.

[Kr] K. Kuratowski, Topology, Vol. 1, Academic Press, N.Y., 1966. MR 36 #840.

[L] J. E. Littlewood, On inequalities in the theory of functions, Proc. London Math. Soc. 23 (1925), 481-519.

[Lu] A. Lubin, Isometries induced by composition operators and invariant subspaces, Illinois J. Math. 19 (1975), 424-427.

[N1] J. von Neumann, Einige Satze über messbare Abbildungen, Ann. Math. (2) 33 (1932), 574-586.

[N2] J. von Neumann, Zur operatoren methode in der klassischen Mechanik, Ann. Math. (2) 33(1932), 587-642, 789-791.

[Nr] E. A. Nordgren, Composition operators, Canad. J. Math. 20 (1968), 442-449. MR 36 #6961.

[R1] W. C. Ridge, Composition operators, Thesis, Indiana U., 1969.

[R2] W. C. Ridge, Spectrum of a composition operator, Proc. Amer. Math. Soc. 37 (1973), 121-127. MR 46 #5583.

[R3] W. C. Ridge, Characterization of abstract composition operators, Proc. Amer. Math. Soc. 45 (1974), 393-396. MR 49 #11310.

[Rn1] R. C. Roan, Composition operators on the space of functions with H^p derivative, to appear.

[Rn2] R. C. Roan, Composition operators on a space of Lipschitz functions, to appear.

[Rn3] R. C. Roan, Composition operators on H^p with dense range, Indiana U. Math. J., to appear.

[Rd1] W. Rudin, Analytic functions of class H^p, Lectures on Functions of a Complex Variable (W. Kaplan ed.), U. Michigan Press, Ann Arbor, 1955. MR 17-24.

[Rd2] W. Rudin, Analytic functions of class H_p, Trans. Amer. Math. Soc. 78 (1955), 46-66. MR 16-810.

[Rf] J. V. Ryff, Subordinate H^p functions, Duke Math. J. 33 (1966), 347-354. MR 33 #289.

[Sa] D. E. Sarason, Weak-star generators of H^∞, Pacific J. Math. 17 (1966), 519-528. MR 35 #2151.

[Sc] H. J. Schwartz, Composition operators on H^p, Thesis, U. Toledo, 1969.

[ST] J. H. Shapiro and P. D. Taylor, Compact, nuclear, and Hilbert-Schmidt composition operators on H^2, Indiana U. Math. J. 23 (1973/74), 471-496. Mr 48 #4816.

[Sh] A. L. Shields, Weighted shift operators and analytic function theory, Topics in Operator Theory, Mathematical Surveys, No. 13, Amer. Math. Soc., Providence, 1974. MR 50 #14341.

[SW] A. L. Shields and L. J. Wallen, The commutants of certain Hilbert space operators, Indiana U. Math. J. 20 (1971), 777-788. MR 44 #4558.

[Sk1] R. Sikorski, On the inducing of homomorphisms by mappings, Fund. Math. 36 (1949), 7-22. MR 11-166.

[Sk2] R. Sikorski, Boolean algebras, 2nd ed., Ergebnisse der Mathematik und ihrer Grenzgebiete, N.F., Band 25, Springer-Verlag, Berlin, 1964. MR 31 #2178.

[Sn1] R. K. Singh, Composition operators, Thesis, U. New Hampshire, 1972.

[Sn2] R. K. Singh, Compact and quasinormal composition operators, Proc. Amer. Math. Soc. 45 (1974), 80-82. MR 50 #1043.

[Sn3] R. K. Singh, Normal and Hermitian composition operators, Proc. Amer. Math. Soc. 47 (1975), 348-350. MR 50 #8153.

[Sn4] R. K. Singh, Invertible composition operators on $L^2(\lambda)$, Proc. Amer. Math. Soc. 56 (1976), 127-129. MR 53 #3776.

[Sn5] R. K. Singh, Composition operators induced by rational functions, Proc. Amer. Math. Soc. 59 (1976), 329-333. MR 54 #5895.

[Sw] D. W. Swanton, Composition operators on $H^p(D)$, Thesis, Northwestern U., 1974.

[Sz] B. Sz.-Nagy, Über die Gesamtheit der charakteristischen Funktionen im Hilbertschen Funktionenraum, Acta Sci. Math. 9 (1937), 166-176.

[T1] G. Targonski, Seminar on functional operators and equations, Springer Lecture Notes #33, Springer-Verlag, Berlin, 1967. MR 36 #744.

[T2] G. Targonski, Linear endomorphisms of function algebras and related functional equations, Indiana U. Math. J. 20 (1970), 579-589. MR 42 #5054.

[W] J. Wolff, Sur l'iteration des fonctions, C. R. Acad. Sci. Paris 182 (1926), 42-43, 200-201.

UNIVERSITY OF NEW HAMPSHIRE
DURHAM, NH

ERGODIC GROUPS OF SUBSTITUTION OPERATORS ASSOCIATED WITH ALGEBRAICALLY MONOTHETIC GROUPS

James R. Brown

1. INTRODUCTION

Since this is a conference on concrete operators, I would like to begin by calling to your attention the operator defined on $L_2[0,1]$ by $Tf(x) = f(x + \sqrt{.5})$, where addition is understood to be modulo one. It is easily verified that T is ergodic; that is, admits no nonconstant invariant functions $f \in L_2[0,1]$. It follows easily [4, p. 34] that all eigenvalues of T are simple and that they form a subgroup of the circle group $K = \{z : |z| = 1\}$ in the complex plane. Since the trigonometric polynomials $f_n(x) = \exp(2\pi inx)$ are eigenfunctions of T, it follows that $L_2[0,1]$ is a direct sum of one-dimensional eigenspaces of T.

It should be noted here that these properties of T are shared by all composition (substitution) operators

(1) $$Tf(x) = f(x + a)$$

on $L_2[0,1]$, provided only that a is irrational. For then $\{na : n \in Z\}$ is dense in the additive group $[0,1]$ (addition mod 1). Ergodicity of T follows easily from this fact. We are now ready to pose (and answer) two questions.

Question 1. What other topological groups have dense subgroups isomorphic to the additive group of integers Z?

Question 2. What other composition operators $Tf(x) = F(\varphi(x))$ have the listed properties?

The groups defined by Question 1 were introduced by D. van Dantzig in 1930 in a footnote of a paper on topological continua [7]. He called them monothetic. The answer to Question 1 was given independently by P. Halmos and H. Samelson [5] in 1942 and by H. Anzai and S. Kakutani [1] in 1943. They showed that G is monothetic if and only if either $G \cong Z$ or G is compact abelian and the dual group

G is isomorphic to an algebraic subgroup of the discrete circle group K_d; that is, K with the discrete topology.

It is further shown in [5] and in [1] that G is monothetic if and only if the cardinality of $\hat{G}$ is no greater than that of K and the torsion subgroup $T(\hat{G})$ of $\hat{G}$ is isomorphic to a subgroup of K_d. This latter condition is equivalent to saying that every finitely generated subgroup of $\hat{G}$ is cyclic ($\hat{G}$ is one-dimensional).

Question 2 was answered by P. Halmos and J. von Neumann in 1942 [6] under the additional assumption that φ is a measure-preserving transformation of some finite measure space X, and T operatoes on $L_2(X)$. Their theorem asserts that every such operator T that is ergodic and has pure point spectrum (the eigenfunctions of T span $L_2(X)$) is unitarily equivalent (this is obvious) to the operator defined by equation (1) on $L_2(G)$, where G is monothetic and $\{na : n \in Z\}$ is dense in G, and moreover, that the equivalence is implemented by another composition operator $Sf(x) = f(\psi(x))$ from $L_2(X)$ to $L_2(G)$ with ψ measure-preserving (not obvious). Thus the answer to Question 1 provides at the same time a complete answer to Question 2 for φ measure-preserving.

2. Generalizations. The concept of an operator T with discrete or pure point spectrum was substantially extended by L. M. Abramov in 1962. Using the notion of generalized eigenvalues and generalized eigenfunctions introduced by Halmos in [4], Abramov defined quasidiscrete spectrum and gave a [6]-type answer to the corresponding Question 2.

While it had apparently not been noticed that Question 2, suitably modified, made sense also when T operates on $C(X)$, X a compact Hausdorff space and φ a homeomorphism of X, and that it has the same answer as in the measure case, this extension was made to the more general work of Abramov by F. J. Hahn and W. Parry in 1965. Again the answer was in the spirit of [6].

Finally, in 1969 the present author incorporated the spirit of Question 1 and showed in both the measure case and the topological case that any T with quasi-discrete spectrum could be represented as a factor of a naturally defined composition operator on the direct product of countably many copies of the Bohr compactification of Z. Details of these results may be found in the monograph [2].

3. Algebraically Monothetic Groups. Another direction of generalization was introduced by the present author in 1972. This was an attempt to obtain for ergodic group automorphisms a representation similar to the one mentioned in the previous paragraph for operators with quasidiscrete spectrum and was touched off by the observation that such an automorphism τ of a compact abelian metric group G has the following property.

Property M. There exists an $a \in G$ such that the orbit of a under powers of τ separates the points of $\hat{G}$.

The link with monothetic groups and discrete spectrum comes from the observation that Property M can be rewritten as follows. Let G be a compact abelian (not necessarily metrizable) group and τ a continuous automorphism of G. Let $\mathcal{E}(\tau)$ denote the group of automorphisms φ of G that can be written

$$\varphi(x) = \sum_j n_j \tau^j(x), \tag{2}$$

where $n_j \in Z$ for each $j \in Z$ and vanishes except for finitely many values of j. The group operation in $\mathcal{E}(\tau)$ is pointwise addition. For each $a \in G$, $\mathcal{E}(\tau)a$ is then the group generated by the orbit of a under τ. If τ is the identity, it reduces to our old friend $\{na : n \in Z\}$.

Property M. There exists an $a \in G$ such that $\mathcal{E}(\tau)a$ is dense in G.

For obvious reasons, we call the system (G,τ,a) monothetic and say that G is algebraically monothetic if there exist such τ and a. Corresponding to Question 1 we have the following.

Question 3. What groups are algebraically monothetic?

In a sense, the complete answer is given in [3] and parallels [5] and [1]. A group G is algebraically monothetic if and only if the cardinality of $\hat{G}$ is no greater than the power of the continuum and $T(\hat{G})$ is isomorphic to a shift-invariant subgroup of K_d^ω. Here K_d^ω denotes the direct product $(-\infty < n < \infty)$ of countably many copies of K_d; that is, the discretized version of the infinite-dimensional torus, and a subgroup H is shift-invariant if $\sigma(H) = H$, where $\sigma(x) = y$ means $y_{n+1} = x_n$.

A potpourri of more specialized, and possibly more useful, results follows.

(A) If G is monothetic, it is algebraically monothetic.

(B) Countable products of monothetic groups are algebraically monothetic.

(C) $G = Z_2 \oplus Z_2$ is algebraically monothetic but not monothetic, as is $\bar{Z} \oplus \bar{Z}$, the Bohr compactification of Z^2.

(D) If G is metric and supports an ergodic τ, then G is algebraically monothetic.

(E) If $\hat{G}$ is finitely generated, then G is algebraically monothetic.

(F) If G is separable and $\hat{G}$ is divisible, then G is algebraically monothetic.

Proofs of these and other facts are contained in [3]. However, the following question remains open.

Question 4. Is every separable compact abelian group algebraically monothetic?

4. The Group $\mathcal{J}$ Generated by $\mathcal{E}(\tau)$. We come finally to the link between algebraically monothetic groups and this conference. What then is the proper reformulation of Question 2? One such formulation follows. There may be others with more interesting answers, and in any case, I don't yet know (for sure) the answer to this one.

For each $\varphi \in \mathcal{E}(\tau)$, define

$$T_\varphi f(x) = f(x + \varphi(a)). \tag{3}$$

Then $\mathcal{J} = \{T_\varphi : \varphi \in \mathcal{E}(\tau)\}$ is a group (under composition) of operators on $L_2(G)$.

Property 1. $\mathcal{J}$ is ergodic.

Proof. For $f \in L_2(G)$ we can write

$$f(x) = \sum_{\gamma \in \hat{G}} (f,\gamma)\, \gamma(x).$$

Then

$$T_\varphi f(x) = \sum_{\gamma\in\hat{G}} (f,\gamma)\gamma(\varphi(a))\gamma(x) .$$

Suppose now that $T_\varphi f = f$ for all $\varphi \in \mathcal{E}(\tau)$. Then

$$(f,\gamma) \neq 0 \Rightarrow \gamma(\varphi(a)) = 1 \quad \text{for all} \quad \varphi \in \mathcal{E}(\tau)$$
$$\Rightarrow \gamma(x) \equiv 1 .$$

Thus f is a constant, and the proof is complete.

Property 2. $\mathcal{J}$ is a factor of Z^∞.

Proof. Clearly, $\mathcal{E}(\tau) \cong Z^\infty$, the direct sum of countably many copies of Z, and $\varphi \to T_\varphi$ is a group homomorphism, $T_{\varphi_1+\varphi_2} = T_{\varphi_1}T_{\varphi_2}$.

Property 3. There is a complete orthonormal basis for $L_2(G)$ consisting of eigenfunctions f of $\mathcal{J}$, that is $Tf = \lambda_{T,f} f$ for each $T \in \mathcal{J}$.

Proof. For each $f = \gamma \in \hat{G}$ and each $\varphi \in \mathcal{E}(\tau)$, we have

(4) $$\lambda_{T_\varphi,\gamma} = \gamma(\varphi(a)) = \prod_j \gamma(a_j)^{n_j} .$$

Property 4. For each eigenfunction f, the mapping $T \to \lambda_{T,f}$ represents $\mathcal{J}$ as a subgroup of K.

Proof. See equation (4).

Question 5. Do Properties 1-4 (or 1-3) characterize groups $\mathcal{J}$ arising in this way among all groups of composition (substitution) operators? If so, is the unitary equivalence given by a composition operator? Can Property 4 be used to construct this operator?

REFERENCES

[1] H. Anzai and S. Kakutani, Bohr compactifications of a locally compact group I, II, Proc. Japan Acad. (Tokyo), 19 (1943), 467-480 and 533-539.

[2] J. R. Brown, Ergodic Theory and Topological Dynamics, Academic Press, New York, 1976.

[3] J. R. Brown, Monothetic automorphisms of a compact abelian group, in Recent Advances in Topological Dynamics, A. Beck [ed], Springer, New York, (1973), 59-77.

[4] P. R. Halmos, Ergodic Theory, Chelsea, New York, 1956.

[5] P. Halmos and H. Samelson, On monothetic groups, Proc. Nat. Acad. Sci. U.S., 28 (1942), 254-258.

[6] P. R. Halmos and J. von Neumann, Operator methods in classical mechanics, II, Ann. of Math. Ser. II, 43 (1942), 332-350.

[7] D. van Dantzig, Über topologisch homogene Kontinua, Funda. Math., 15 (1930), 102-125.

OREGON STATE UNIVERSITY

COMMUTANTS OF ANALYTIC TOEPLITZ OPERATORS WITH AUTOMORPHIC SYMBOL

Carl C. Cowen

For a function φ in H^∞ of the unit disk D, the operator on the Hardy space H^2 of multiplication by φ will be denoted by T_φ and its commutant by $\{T_\varphi\}'$. We consider the special case when φ is a covering map (in the sense of Riemann surfaces) of D onto a bounded plane domain. For such a function φ, let G be the group of linear fractional transformations I, of D onto D such that $\varphi \circ I = \varphi$. ($G$ is isomorphic to the fundamental group of $\varphi(D)$.) For our purposes it will be sufficient to think of an annulus of inner radius r and outer radius R, $0 \leq r < R < \infty$, and the mapping φ obtained by mapping the disk onto the strip $\log r < \operatorname{Re} z < \log R$ followed by the exponential map. The group G in the disk corresponds to the group of translations by $2\pi mi$ in the strip.

We can characterize the operators that commute with T_φ.

THEOREM A. If S is in $\{T_\varphi\}'$, then there are unique analytic functions b_I, for I in G, defined on D so that for each α in D and h in H^2

$$(**) \qquad (Sh)(\alpha) = \sum_{I \in G} b_I(\alpha)\, h(I(\alpha))$$

where the series converges absolutely for each α and uniformly on compact subsets of the disk.

Conversely, if S is a bounded operator that has a representation (**), then S is in $\{T_\varphi\}'$.

COROLLARY B. The following are equivalent:

(i) $\{T_\varphi\}' = \{T_B\}'$ for some inner function B.

(ii) there is a one-to-one analytic function σ so that $\varphi = \sigma \circ B$.

(iii) the union of $\varphi(D)$ and a set of capacity zero is simply connected.

COROLLARY C. T_φ does not commute with any non-zero compact operators.

The condition (**) may be a bit easier to work with if we rewrite it. For I

in G, we define the operator U_I on H^2 by $(U_I h)(\alpha) = \sqrt{I'(\alpha)}\ h(I(\alpha))$ for h in H^2, α in D. Each U_I is unitary, in fact, a bilateral shift of infinite multiplicity [4] and the correspondence $I \to U_I$ is a unitary representation of G. If we define analytic functions c_I on D by $c_I(\alpha) = b_I(\alpha)/\sqrt{I'(\alpha)}$, then (**) becomes $(Sh)(\alpha) = \Sigma\ c_I(\alpha)(U_I h)(\alpha)$. That is, to each S in $\{T_\varphi\}'$ we associate a formal series

$$(*) \qquad S \sim \sum_{I \in G} c_I U_I$$

where the precise meaning is given in the formula (**).

Upon writing (*) questions begin to arise.

<u>Question</u> 1. When does $\Sigma c_I U_I$ represent a bounded operator? In particular, if S is bounded, are the coefficients bounded analytic functions on D?

Obviously, if $\{c_I\} \subset H^\infty$ and $\Sigma \|c_I\|_\infty < \infty$ then $S = \Sigma\ c_I U_I$ is bounded and $\|A\| \leq \Sigma \|c_I\|_\infty$. If $\{a_n\}_{-\infty}^{\infty}$ is a sequence of scalars so that $\Sigma_{-\infty}^{\infty}\ a_n\ e^{in\theta} \in L^\infty$, and J is a fixed element of G, then $\Sigma_{-\infty}^{\infty}\ a_n\ U_{J^n}$ represents a bounded operator. (Proof: Functional calculus for U_J.)

THEOREM D. If S in $\{T_\varphi\}'$ has representation (*) and F is a finite subset of G so that $c_I \equiv 0$ for $I \notin F$, then $c_I \in H^\infty$ for I in F and $\|c_I\|_\infty \leq M_F\|S\|$, where M_F is a constant depending only on F.

<u>Question</u> 1'. Is the set of operators for which all but finitely many of the c_I are zero dense in $\{T_\varphi\}'$?

In some cases, the coefficients need not be bounded functions. Let $\varphi(z) = [\exp((z+1)(z-1)^{-1}) - e^{-1}]\ [1 - e^{-1}\exp((z+1)(z-1)^{-1})]^{-1}$. Then φ is a covering map of the disk onto the disk punctured at $-e^{-1}$, and the covering group for φ is generated by a single linear fractional transformation, J. In fact, φ is a Blaschke product with zeroes at $\ldots, J^{-1}(J^{-1}(0)), J^{-1}(0), 0, J(0), J(J(0)), \ldots$. Now let B be the Blaschke product with zeroes at $J(0), J(J(0)), \ldots$ and let P be the orthogonal projection of H^2 onto the subspace spanned by $B, B\varphi, B\varphi^2, B\varphi^3, \ldots$. It is easy to see that P commutes with T_φ and a somewhat tedious calculation yields an explicit formula for the coefficients in the series for P. For example, the coefficient c_0, corresponding to the identity of the group, is

$$c_0(z) = \varphi(z)[z\varphi'(z)]^{-1}.$$

Since $c_0(x)$ approaches infinity as x approaches 1 along the real axis, c_0 is not bounded in D.

We say an operator S in $\{T_\varphi\}'$ <u>lifts</u> if there is an operator $\tilde{S}$ on L^2 so that $\tilde{S}|_{H^2} = S$ and $\tilde{S}$ commutes with M_φ, multiplication by φ in L^2. (The general question of lifting commutants of analytic Toeplitz operators is open.) For <u>each I</u> in G, U_I has an obvious lifting: define $\tilde{U}_I$ on L^2 by $(\tilde{U}_I h)(e^{i\theta}) = \sqrt{I'(e^{i\theta})}\, h(I(e^{i\theta}))$.

<u>Question</u> 2. If $S \sim \sum c_I U_I$ is a bounded operator on H^2, does $\sum c_I \tilde{U}_I$ represent a bounded operator on L^2?

In the previously mentioned cases for which S is obviously bounded, $\sum \|c_I\|_\infty < \infty$ or $S \sim \sum_{-\infty}^{\infty} a_n U_j^n$ for constants a_n with $\sum_{-\infty}^{\infty} a_n e^{in\theta}$ in L^∞, the answer is obviously yes. We also have a more general result.

THEOREM E. If $S \sim \sum c_I U_I$ and H is a finitely generated subgroup of G such that $c_I \equiv 0$ for $I \notin H$, then S has a lifting $\tilde{S}$ and $\|S\| = \|\tilde{S}\|$.

<u>Proof</u>. Let Ω be the Riemann surface obtained from the disk and the subgroup H by identifying a point of the disk with each of its images under elements of H; thus, the universal covering space of Ω is the disk and H is its covering group. Let π be the covering projection of D onto Ω. Let ψ be an Ahlfors map of Ω into the disk. Since H is finitely generated, Ω is finitely connected and $\psi \circ \pi$ is an inner function. From the form of S, we see that S commutes with $T_{\psi\circ\pi}$. Now $\{T_{\psi\circ\pi}\}'$ lifts, so S has a lifting $\tilde{S}$ which commutes with $M_{\psi\circ\pi}$ (and $\|S\| = \|\tilde{S}\|$). To see that $\tilde{S}$ commutes with M_φ, we observe that $M_\varphi\tilde{S} - \tilde{S}M_\varphi$ commutes with $M_{\psi\circ\pi}$ and that $M_\varphi\tilde{S} - \tilde{S}M_\varphi$ restricted to H^2 is $T_\varphi S - ST_\varphi = 0$. Since the lifting of an operator in the commutant of a subnormal operator to the commutant of the minimal normal extension is unique if it exists, [1], and since the zero operator on L^2 is an obvious lifting of the zero operator on H^2, we

have $M_\varphi\tilde{S} - \tilde{S}M_\varphi = 0$.

<u>Question</u> 3. Which operators in $\{T_\varphi\}'$ are self adjoint, especially which are self adjoint projections?

If $S \sim \sum c_I U_I$, we may write formally $S^* \sim \sum U_I^* \overline{c_I} = \overline{\sum c_I \circ I^{-1}}\, U_{I^{-1}}$ and we might expect that if $S = S^*$ that the functions c_I would be forced to be constant. This is not the case though; the difficulty is that the representation for S^* is not in the form required to conclude uniqueness of coefficients.

Suppose φ is a covering map of an annulus or punctured disk so that the group G is singly generated, by J, and let $U = U_J$. If $S \sim \sum_{-\infty}^{\infty} a_n U^n$, where the a_n are constants, then $S = S^*$ if and only if $\sum_{-\infty}^{\infty} a_n e^{in\theta}$ is a real-valued L^∞ function. If these were all the self adjoint operators in $\{T_\varphi\}'$, the lattice of reducing subspaces for $\{T_\varphi\}'$ would be isomorphic to the lattice of measurable subsets of ∂D. But if φ covers a punctured disk, φ is inner and the lattice of reducing subspaces for T_φ is isomorphic to the lattice of subspaces of H^2. (If φ covers an annulus, it is known that no other reducing subspaces exist.)

However, we do have the following positive result.

THEOREM F. Let φ be a covering map whose group G is generated by the linear fractional transformation J. The self adjoint operators which commute with both T_φ and U_J are the operators $\sum_{-\infty}^{\infty} a_n U_J^n$ where $\sum_{-\infty}^{\infty} a_n e^{in\theta}$ is the Fourier series of a real-valued L^∞ function.

COROLLARY G. The lattice of common reducing subspaces for T_φ and U_J is isomorphic to the lattice of measurable subsets of ∂D.

(Let K_α be the kernel function for evaluation of H^2 functions at the point α of the disk, that is, $K_\alpha(z) = (1 - \bar{\alpha}z)^{-1}$.)

<u>Proof</u>. Without loss of generality, we assume $\varphi(0) = 0$. We will write U for U_J and c_n for the function c_{J^n}.

Suppose A is a self-adjoint operator in $\{T_\varphi\}' \cap \{U\}'$, and $A \sim \sum c_n U^n$. We have $UA \sim \sum c_n \circ J\, U^{n+1}$ and $AU \sim \sum c_n U^{n+1}$, so by the uniqueness of the coefficients, $c_n \circ J = c_n$. Since J generates the covering group for φ, this means that each c_n is of the form $b_n \circ \varphi$ for some analytic function b_n.

If $\varphi(\alpha) = 0$, then $AK_\alpha = A^*K_\alpha = \sum U^{-n}(\overline{c_n(\alpha)}\, K_\alpha) = \overline{\sum b_n(0)}\, U^{-n} K_\alpha$. Since the closed span of $\{K_\alpha \mid \varphi(\alpha) = 0\}$ is the kernel of T_φ^*, we have A restricted to $\ker(T_\varphi^*)$ (an invariant subspace for A) is equal to $\sum_{-\infty}^{\infty} b_n(0)\, U^n$ restricted to $\ker(T_\varphi^*)$. But U restricted to $\ker(T_\varphi^*)$ is unitarily equivalent to the bilateral shift (of multiplicity one) so $\sum b_n(0)\, e^{in\theta}$ is the Fourier series of an L^∞ function, and $B = \sum_{-\infty}^{\infty} b_n(0)\, U^n$ is bounded on all of H^2.

Thus $A - B$ is a self-adjoint operator on H^2 which commutes with U and T_φ. By construction $A - B = 0$ on $\ker(T_\varphi^*)$, so if h is in $\ker(T_\varphi^*)$, $(A - B)(\varphi^n h) = T_\varphi^n(A - B)(h) = 0$, for $n = 0,1,2,\ldots$. Since H^2 is spanned by $\{\varphi^n h \mid n \in \ker(T_\varphi^*), n = 0,1,2,\ldots\}$ we see that $A - B = 0$ and $A = \sum_{-\infty}^{\infty} b_n(0)\, U^n$.

It is clear that the above results depend ultimately on the interaction between the analytic Toeplitz operator and some composition operators. Further study of relationships between Toeplitz operators and composition operators will undoubtedly reveal much about the structure of each.

REFERENCES

[1] J. Bram, Subnormal operators, Duke Math. J., 22 (1955), 75-94.

[2] C. Cowen, The commutant of an analytic Toeplitz operator, to appear, Trans. Amer. Math. Soc.

[3] C. Cowen, The commutant of an analytic Toeplitz operator, II, to appear.

[4] A. Lubin, Isometries induced by composition operators and invariant subspaces, Illinois J. Math., 19 (1975), 424-427.

UNIVERSITY OF ILLINOIS AT URBANA-CHAMPAIGN

ANOTHER DESCRIPTION OF NEST ALGEBRAS

James A. Deddens

Given a nest $\mathcal{N}$ of subspaces on a separable Hilbert space $\mathcal{H}$, Ringrose [9] introduced the concept of the associated nest algebra $\mathcal{A}_{\mathcal{N}}$ defined by $\mathcal{A}_{\mathcal{N}} = \{X \in \mathcal{B}(\mathcal{H}) : \text{every } N \in \mathcal{N} \text{ is invariant for } X\}$. Nest algebras have been studied by many other authors since; e.g., [1], [3], [5], [6], [7], [10]. In this article we will study the following set and its relationship to nest algebras: given an invertible operator A, let

$$\mathcal{B}_A = \{X \in \mathcal{B}(\mathcal{H}) : \|A^k X A^{-k}\| \text{ bounded for } k = 0,1,2,3,\ldots\}.$$

For $X \in \mathcal{B}_A$ let $c_A(X) = \sup_{0 \leq k} \|A^k X A^{-k}\| < +\infty$. In general $\mathcal{B}_A$ is an algebra that contains the commutant $\{A\}'$ of A. We will show that if A is a positive invertible operator, then $\mathcal{B}_A$ is equal to the nest algebra associated with the nest $\{E([0,a]) : a \geq 0\}$, where E is the spectral measure for A, and that $c_A(X) = \|X\|$. Conversely every nest algebra arises in this manner. The algebra $\mathcal{B}_A$ has been studied previously:

THEOREM 0 [2]. If $A \in \mathcal{B}(\mathcal{H})$ is of the form $A = bI + N$ where $0 \neq b \in \mathbb{C}$ and N is nilpotent, then $\mathcal{B}_A = \{A\}'$. Furthermore, if $\mathcal{H}$ is finite dimensional, then the converse holds.

In this paper we are more concerned with the opposite extreme: when $\mathcal{B}_A$ is much larger than $\{A\}'$. In case $\mathcal{H}$ is finite dimensional, it is possible to completely describe $\mathcal{B}_A$ for any A. Since the formulation is somewhat complicated we shall omit it. It almost follows from Theorem 0, Lemma 1 and Proposition 2.

We now give some notation and definitions.

$\mathcal{H}$ will denote a complex separable Hilbert space, and $\mathcal{B}(\mathcal{H})$ will denote the algebra of all bounded linear operators on $\mathcal{H}$. A family $\mathcal{N}$ of subspaces of $\mathcal{H}$ is called a <u>nest</u> if it is totally ordered by inclusion. A nest $\mathcal{N}$ is complete if (i) $\{0\}, \mathcal{H} \in \mathcal{N}$ and (ii) for

(i) $\{0\}, \mathcal{H} \in \mathfrak{n}$ and (ii) for any nonempty subset $\mathfrak{n}_0 \subseteq \mathfrak{n}$, the subspaces $\wedge\{N : N \in \mathfrak{n}_0\}$ and $\vee\{N : N \in \mathfrak{n}_0\}$ are both members of $\mathfrak{n}$ can be completed, and Lemma 3.2 of [9] shows that the corresponding nest algebras are equal. For this reason, it is convenient to work with complete nests whenever possible. The core $\mathcal{C}_{\mathfrak{n}}$ of a complete nest $\mathfrak{n}$ is the von Neumann algebra generated by the self-adjoint projections onto the members of $\mathfrak{n}$. The diagonal $\mathcal{D}_{\mathfrak{n}}$ of a nest $\mathfrak{n}$ is the von Neumann algebra $\mathcal{D}_{\mathfrak{n}} = \mathcal{C}_{\mathfrak{n}} \cap \mathcal{C}_{\mathfrak{n}}^* \subseteq \mathcal{C}_{\mathfrak{n}}$ and $\mathcal{D}_{\mathfrak{n}} = \mathcal{C}_{\mathfrak{n}}'$.

There are several different descriptions of nest algebras in the literature. R. Loebl and P. Muhly [6] show that nest algebras are precisely the algebras of analytic operators with respect to certain groups of inner *-automorphisms. (We take this opportunity to point out that Loebl and Muhly use the nest $E([a,\infty))$: $a \geq 0$)). W. Longstaff [7] uses the structure theory of nests developed by J. Erdos [3] to show that every nest algebra is the weakly closed algebra generated by two operators D and B on a direct sum of L^2-spaces. Finally, J. Schue [10] describes hyperreducible triangular algebras (i.e., nest algebras for which the diagonal is a maximal abelian algebra) as the algebraic sum of the diagonal $\mathcal{D}_{\mathfrak{n}} = \{A\}'$ and a two-sided ideal defined in terms of certain boundedness conditions involving the operator D_A, where $D_A(X) = AX - XA$. We will give another description of nest algebras solely in terms of the operator A.

LEMMA 1. $\mathcal{B}_A$ is an algebra that contains the commutant $\{A\}'$ of A. If $A = TA_1T^{-1}$ then $\mathcal{B}_A = \mathcal{B}_{TA_1T^{-1}} = T\,\mathcal{B}_{A_1}T^{-1}$. Also $\mathcal{B}_A^* = \mathcal{B}_{A^{*-1}}$.

Proof. $\mathcal{B}_A$ clearly contains $\{A\}'$ and is an algebra because $\|A^kX_1X_2A^{-k}\| = \|A^kX_1A^{-k}A^kX_2A^{-k}\| \leq \|A^kX_1A^{-k}\|\,\|A^kX_2A^{-k}\|$. Hence $c_A(X_1X_2) \leq c_A(X_1)c_A(X_2)$. Next

$$\|(TA_1T^{-1})^kX(TA_1T^{-1})^{-k}\| = \|TA_1^k\,T^{-1}XTA_1^{-k}\,T^{-1}\| \leq \|T\|\,\|T^{-1}\|\,\|A_1^k\,T^{-1}XTA_1^{-1}\|,$$

thus $T\,\mathcal{B}_AT^{-1} \subseteq \mathcal{B}_{TA_1T^{-1}}$; also

$$\|A_1^k T^{-1}XTA_1^{-k}\| = \|T^{-1}(TA_1T^{-1})^k X(TA_1T^{-1})^{-k}T\| \leq \|T\|\ \|T^{-1}\|\ \|(TA_1T^{-1})^k\ X(TA_1T^{-1})^{-k}\|;$$

thus $\mathcal{B}_{TA_1T^{-1}} \subseteq T\,\mathcal{B}_A T^{-1}$. The final statement follows since $\|A^kXA^{-k}\| = \|(A^kXA^{-k})^*\| = \|A^{*-k}X^*A^{*k}\|$.

PROPOSITION 2. Suppose A is an invertible operator on $\bigoplus_{i=1}^{n} \mathcal{H}$ of the form $A = A_1 \oplus A_2 \oplus \cdots \oplus A_n$. If $\|A_i\|\ \|A_{i+1}^{-1}\| < 1$ for $i = 1,2,\ldots,n-1$, then

$$\mathcal{B}_A = \left\{ X = (X_{ij}) \in \mathcal{B}\left(\bigoplus_{i=1}^{n} \mathcal{H}\right) : X_{ii} \in \mathcal{B}_{A_i} \text{ and } X_{ij} = 0 \text{ whenever } i > j \right\}.$$

That is, $\mathcal{B}_A$ is all upper triangular matrices whose diagonal entries belong to the corresponding $\mathcal{B}_{A_i}$.

<u>Proof</u>. Suppose $X = (X_{ij}) \in \mathcal{B}\left(\bigoplus_{i=1}^{n} \mathcal{H}\right)$. Then $(A^kXA^{-k})_{ij} = A_i^kX_{ij}A_j^{-k}$. If $X \in \mathcal{B}_A$ then clearly $X_{ii} \in \mathcal{B}_{A_i}$. The hypothesis $\|A_i\|\ \|A_{i+1}^{-1}\| < 1$ implies that $\|A_i\|\ \|A_j^{-1}\| < 1$ whenever $i < j$, since

$$\|A_i\|\ \|A_j^{-1}\| \leq (\|A_i\|\ \|A_{i+1}^{-1}\|)(\|A_{i+1}\|\ \|A_{i+2}^{-1}\|) \cdots (\|A_{j-1}\|\ \|A_j^{-1}\|).$$

Now suppose $X \in \mathcal{B}_A$, $i > j$ and $X_{ij} \neq 0$. Then there exists $p \in \mathcal{H}$ with $X_{ij}p = q \neq 0$. Since A_j^{-k} is invertible there exist $p_k \in \mathcal{H}$ with $A_j^{-k}p_k = p$. Then

$$\left\| A_i^kX_{ij}A_j^{-k} \frac{p_k}{\|p_k\|} \right\| = \|A_i^kq\|/\|A_j^kp\| \geq \frac{\|q\|}{\|A_i^{-1}\|^k} \frac{1}{\|A_j\|^k\|p\|}$$

$$= \frac{\|q\|/\|p\|}{(\|A_i^{-1}\|\|A_j\|)^k} \to +\infty \quad \text{as} \quad k \to +\infty.$$

But this implies that $\|A_i^kX_{ij}A_j^{-k}\| \to +\infty$ as $k \to +\infty$, contradicting $x \in \mathcal{B}_A$. Hence $X_{ij} = 0$ for $i > j$ whenever $X \in \mathcal{B}_A$.

On the other hand suppose $X \in \mathcal{B}\left(\bigoplus_{i=1}^{n} \mathcal{H}\right)$ satisfies $X_{ij} \in \mathcal{B}_A$ and $X_{ij} = 0$

whenever $i > j$. We need only show $\{\|(A^k X A^{-k})_{ij}\|\}_{k=0}^{\infty}$ is bounded for all $1 \leq i,\ j \leq n$. This is clear for $i \geq j$. Suppose $i < j$. Then

$$\|A_i^k X_{ij} A_j^{-k}\| \leq \|A_i\|^k \ \|X_{ij}\| \ \|A_j^{-1}\|^k \leq (\|A_i\| \ \|A_j^{-1}\|)^k \ \|X_{ij}\| \to 0$$

$$\text{as } k \to 0 .$$

Hence $X \in \mathcal{B}_A$.

PROPOSITION 3. There is a projection π of $\mathcal{B}_A$ onto $\{A\}'$ that satisfies $\pi(A_1 X A_2) = A_1 \pi(X) A_2$ for $A_1, A_2 \in \{A\}'$. If there is a constant C such that $c_A(X) \leq C\|X\|$ for all $X \in \mathcal{B}_A$, then π is bounded in norm by C.

Proof. Let glim be a fixed "Banach generalized limit" on ℓ^{∞}. For $X \in \mathcal{B}_A$ and $f,g \in \mathcal{H}$ define $\pi(X)$ by

$$(\pi(X)f,g) = \operatorname{glim}(A^k X A^{-k} f,g) .$$

Then $\pi(X) \in \mathcal{B}(\mathcal{H})$ and $\|\pi(X)\| \leq c_A(X)$. It is clear that $\pi(A_1) = A_1$ for all $A_1 \in \{A\}'$. Now

$$\begin{aligned}(\pi(X)A^{-1}f, A^*g) &= \operatorname{glim}(A^k X A^{-k} A^{-1} f, A^* g)\\ &= \operatorname{glim}(A^{k+1} X A^{-(k+1)} f, g)\\ &= (\pi(X)f,g) .\end{aligned}$$

Hence $A\pi(X)A^{-1} = \pi(X)$. Thus $A\pi(X) = \pi(X)A$ and $\pi(X) \in \{A\}'$ for all $x \in \mathcal{B}_A$. Finally, if $A_1, A_2 \in \{A\}'$ then

$$\begin{aligned}(\pi(A_1 X A_2)f,g) &= \operatorname{glim}(A^k A_1 X A_2 A^{-k} f,g)\\ &= \operatorname{glim}(A^k X A^{-k} A_2 f, A_1^* g)\\ &= (\pi(X)A_2 f, A_1^* g) .\end{aligned}$$

Hence $\pi(A_1 X A_2) = A_1 \pi(X) A_2$.

LEMMA 4. (a) If $A = U|A|$ is an invertible normal operator, then $\mathcal{B}_A = \mathcal{B}_{|A|}$,

(b) If A is an invertible positive operator with spectral measure E, then

$$\mathcal{B}_A \subseteq \mathrm{Alg}(\{E([0,\lambda])\mathcal{H} : \lambda \geq 0\}),$$

(c) If A is an invertible positive operator, then $\mathcal{B}_{A^{-1}} = \mathcal{B}_A^*$ and hence $\mathcal{B}_A \cap \mathcal{B}_A^* = \mathcal{B}_A \cap \mathcal{B}_{A^{-1}} = \{A\}'$.

Proof. (a) Since A is normal, $A^k = (U|A|)^k = U^k|A|^k$; thus

$$\|A^kXA^{-k}\| = \|U^k|A|^kX|A|^{-k}U^{-k}\| = \||A|^kX|A|^{-k}\|.$$

(b) Recall [5] that if E is the spectral measure for positive A, then

$$E([0,\lambda])\mathcal{H} = \{f \in \mathcal{H} : \exists\ C > 0\ \|A^kf\| \leq \lambda^kC\ k = 0,1,2,\ldots\}.$$

If $f \in E([0,\lambda])\mathcal{H}$, and if $X \in \mathcal{B}_A$, then one has

$$\|A^kXf\| = \|A^kXA^{-k}A^kf\| \leq \|A^kXA^{-k}\|\ \|A^kf\| \leq c_A(X)\ \lambda^k\ C = C'\ \lambda^k.$$

So $Xf \in E([0,\lambda])\mathcal{H}$.

(c) It follows from Lemma 1 that $\mathcal{B}_{A^{-1}} = \mathcal{B}_A^*$. If $x \in \mathcal{B}_A \cap \mathcal{B}_{A^{-1}}$ then part (b) implies that X commutes with all the projections $E([0,\lambda])$, and it follows that X commutes with A.

THEOREM 5. Let A be a positive invertible operator and let $\mathcal{N}(A)$ be the completion of the nest $\{E([0,a]) : a \geq 0\}$. Then $\mathcal{B}_A = \mathcal{A}_{\mathcal{N}(A)}$. Conversely, if $\mathcal{N}$ is any complete nest of subspaces, then there exists a positive invertible operator A with $\mathcal{B}_A = \mathcal{A}_{\mathcal{N}}$. Furthermore, for $X \in \mathcal{B}_A$ one has that $\|A^kXA^{-k}\| \leq \|X\|$ for all $k = 0,1,2,\ldots$.

Proof. By Lemma 1 (b) we know that $\mathcal{B}_A \subseteq \mathcal{A}_{\mathcal{N}(A)}$. To see the reverse inclusion we use Theorem IV.2.3 of [6]. Let $\{\alpha_t\}_{t\in\mathbb{R}}$ be the group of inner *-automorphisms of $\mathcal{B}(\mathcal{H})$ given by

$$\alpha_t(X) = A^{-it}XA^{it}.$$

Then Theorem IV.2.3 states that the nest algebra $\mathcal{A}_{\mathcal{N}(A)}$ is equal to the algebra of analytic operators with respect to $\{\alpha_t\}_{t\in\mathbb{R}}$. That is, if $X \in \mathcal{A}_{\mathcal{N}(A)}$ then $\alpha_t(X)$ admits a bounded analytic extension to the upper half-plane, namely, $\alpha_z(X) = A^{-iz}XA^{iz}$, which must be bounded by $\sup_{t\in\mathbb{R}} \|A^{-it}XA^{it}\| = \|X\|$. By merely evaluating $\alpha_z(X)$ at the points $z = 0, i, 2i, 3i, \ldots$, one sees that $\|A^kXA^{-k}\|$ is bounded by $\|X\|$ for $k = 0,1,2,3,\ldots$. Thus $X \in \mathcal{B}_A$.

Conversely if $\mathcal{N}$ is any complete nest of subspaces, one can construct (for example, see Proposition IV.2.1, in [6]) a spectral measure E on $\mathbb{R}$ whose closed support is a closed subset Ω of $[0,1]$, and an order preserving bijection $a \mapsto N_a$ between Ω and $\mathcal{N}$ such that $N_a = E([0,a])\mathcal{H}$. If A equals $\int_0^1 \exp(a)\, dE(a)$ then A is an invertible positive operator, with $\mathcal{N}(A) = \mathcal{N}$. Hence $\mathcal{B}_A = \mathcal{A}_{\mathcal{N}(A)} = \mathcal{A}_{\mathcal{N}}$.

REMARK 6. The preceeding proof uses the deep result, Theorem IV.2.3 of [6]. It is possible to give a more elementary proof of Theorem 5 based on the structure theorems of [3], [7]. However, this argument only yields that the weak closure of $\mathcal{B}_A$ equals $\mathcal{A}_{\mathcal{N}(A)}$. We would like to find an elementary proof that $\mathcal{B}_A$ is weakly closed in case A is positive. We will now outline this alternate proof, since it involves concrete multiplication and integration operators. Given an invertible positive operator A, form the nest $\mathcal{N}(A)$. By [7] we can consider $\mathcal{A}_{\mathcal{N}(A)}$ to be an algebra of operators on a direct sum of L^2-spaces, which is weakly generated by two operators, one of which D is a multiplication operator and the other operator B has an operator matrix whose entries are adjoints of Volterra operators. Since in this representation A is also a multiplication operator by an increasing positive function, it follows that D commutes with A, and hence belongs to $\mathcal{B}_A$. The fact that B belongs to $\mathcal{B}_A$ follows from an elementary calculation analogous to the following: if g is an increasing positive function on $[0,1]$ and $0 \le f \in L^2(\mu)$, then $g^k(x) \int_{[x,1]} f(t)\, g^{-k}(t)\, d\mu(t) \le \int_{[x,1]} f(t)\, d\mu(t)$ for almost all x in $[0,1]$ and all $k = 0,1,2,3,\ldots$. Hence if M_g denotes multiplication by g and V denotes definite integration, then $M_g^k V^* M_g^{-k} f \le V^* f$ for $f \ge 0$. It follows that $V^* \in \mathcal{B}_{M_g}$. Likewise, in the terminology above, B belongs to $\mathcal{B}_A$. Hence, weak closure of $\mathcal{B}_A = \mathcal{A}_{\mathcal{N}(A)}$.

REMARK 7. In the notation of Theorem 5, it is easy to see that $\{A\}'$ equals the diagonal algebra $\mathcal{D}_{\mathfrak{n}(A)}$, while the von Neumann algebra generated by A, $W^*(A)$, equals the core $\mathcal{C}_{\mathfrak{n}(A)}$. The projection given by Proposition 3 has norm 1, and corresponds to the "diagonal map" studied by D. Larson [5]. Also purely atomic nests [5] correspond to diagonal operators A.

The radical $\mathcal{R}$ of any complex normed algebra $\mathfrak{A}$ with identity is [8]:

$$\mathcal{R} = \{T \in \mathfrak{A};\ TS \text{ is quasinilpotent for all } S \in \mathfrak{A}\}.$$

For an invertible operator A let

$$\mathcal{R}_A = \{X \in \mathcal{B}(\mathcal{H}) : \|A^k X A^{-k}\| \to 0 \text{ as } k \to \infty\}.$$

LEMMA 8. $\mathcal{R}_A$ is a two-sided ideal in $\mathcal{B}_A$ that is contained in the radical of $\mathcal{B}_A$.

Proof. Clearly $\mathcal{R}_A$ is a two-sided ideal in $\mathcal{B}_A$. If $T \in \mathcal{R}_A$ and $S \in \mathcal{B}_A$ then

$$\|A^k TS A^{-k}\| \le \|A^k T A^{-k}\| \cdot \|A^k S A^{-k}\| \to 0 \text{ as } k \to \infty.$$

Hence, if $r(X)$ = spectral radius of X, then

$$r(TS) = r(A^k TS A^{-k}) \le \|A^k TS A^{-k}\| \to 0.$$

Thus TS is quasinilpotent.

REMARK 9. Unfortunately $\mathcal{R}_A$ is not always the radical of $\mathcal{B}_A$, even for positive A. This can be seen using Ringrose's [9] characterization of the radical of the nest algebra $\mathfrak{A}_{\mathfrak{n}(A)}$. For if a is a two-sided limit point of the spectrum of A then it is easy to construct operators X such that $Y = E([0,a])\, X[I - E([0,a])]$ belongs to the radical of the next algebra $\mathfrak{A}_{\mathfrak{n}(A)} = \mathcal{B}_A$ but does not belong to $\mathcal{R}_A$.

OPEN QUESTIONS AND CONCLUDING REMARKS

It follows from Theorem 5 that for positive invertible operators A, $\mathcal{B}_A$ is weakly closed and that $c_A(X) = \|X\|$. It turns out that neither of these results hold for arbitrary invertible operators A. For example, let A be the bilateral weighted shift with weight sequence: $(\ldots,1,1,1,2,2,2,\ldots)$; i.e., A acts on ℓ^2 by $Ae_n = 2e_{n+1}$ for $n \geq 0$ and $Ae_n = e_{n+1}$ for $n < 0$. A simple calculation shows that $c_A(e_0 \otimes e_{-n}) = 2^n\|e_0 \otimes e_{-n}\| = 2^n$, for $n = 0,1,2,3,\ldots$ where $e_0 \otimes e_{-n}$ denotes the rank 1 operator defined by $(e_0 \otimes e_{-n})(f) = (f,e_{-n})e_0$. Thus there does not exist a constant C such that $c_A(X) \leq C\|X\|$ for all $X \in \mathcal{B}_A$. By taking appropriate linear combinations of the operators $e_0 \otimes e_{-n}$ one can show that $\mathcal{B}_A$ is not even norm closed.

<u>Question</u> 1. If $\mathcal{H}$ is infinite dimensional, when is $\mathcal{B}_A = \{A\}'$?

REMARK. We would conjecture that $\mathcal{B}_A = \{A\}'$ if A is quasi-similar to an operator of the form $aI + Q$ where $0 \neq a \in \mathbb{C}$ and Q is a quasinilpotent operator. Note that if $\mathcal{B}_A = \{A\}'$ and if A is quasi-similar to C (that is, there exist operators X and Y which are one-to-one, have dense range, and satisfy $AX = XC$ and $YA = CY$), then $\mathcal{B}_C = \{C\}'$. For if $S \in \mathcal{B}_C$, then

$$\|A^k XSY A^{-k}\| = \|X C^k S C^{-k} Y\| \leq \|X\| \cdot \|C^k S C^{-k}\| \cdot \|Y\| .$$

Hence $XSY \in \mathcal{B}_A = \{A\}'$, and we obtain $XSYA = AXSY$. Thus $XSCY = XCSY$. Since X is one-to-one and Y has dense range, we conclude that $SC = CS$ or $S \in \{C\}'$.

It would be of interest to determine whether or not $A = aI + U_+$ satisfies $\mathcal{B}_A = \{A\}'$ when $1 < |a|$ and U_+ is the unilateral shift.

<u>Question</u> 2. Suppose A is not similar to a scalar multiple of a unitary operator. Can $\mathcal{B}_A$ ever equal $\mathcal{B}(\mathcal{H})$? Or ever contain $\mathcal{K}(\mathcal{H})$?

REMARK. The answer to Question 2 is negative in case $\mathcal{H}$ is finite dimensional. The obvious way for $\mathcal{B}_A$ to be equal to all of $\mathcal{B}(\mathcal{H})$ is for $\|A^k\| \cdot \|A^{-k}\|$ to be bounded. We conclude by showing that this implies that A is similar to a scalar multiple of a unitary operator. Suppose $\|A^k\| \cdot \|A^{-k}\| \leq M$ for $k = 0,1,2,3,\ldots$. Taking k^{th} roots and letting $k \to \infty$, we see that $r(A)r(A^{-1}) \leq 1$ where $r(A)$ denotes the spectral radius of A. Since the reverse inequality is trivial, we see that $r(A)r(A^{-1}) = 1$. Hence the spectrum of A lies on a circle centered at the origin. Suppose the radius is a. Then $A_1 = A/a$ has spectrum on the unit circle, and satisfies $\|A_1^k\| \cdot \|A_1^{-k}\| = \|A^k\| \cdot \|A^{-k}\| \le M$ for $k = 0,1,2,3,\ldots$. Since A_1^{-1} also has spectrum contained in the unit circle, we have that $1 \le \|A_1\|, \|A_1^{-1}\|$. Hence

$$\|A_1^k\| \le \|A_1^k\| \cdot \|A_1^{-k}\| \le M$$

for all $k = 0, \pm 1, \pm 2, \ldots$. It follows from a theorem of R. Sz.-Nagy [11], that A_1 is similar to a unitary operator and we obtain the desired conclusion.

REFERENCES

[1] W. Arveson, Interpolation problems in nest algebras, J. Functional Anal., 20 (1975), 208-233.

[2] J. A. Deddens and T. K. Wong, The commutant of analytic Toeplitz operators, Trans. Amer. Math. Soc., 184 (1973), 261-273.

[3] J. A. Erdos, Unitary invariants for nests, Pacific J. Math., 23 (1967), 229-256.

[4] P. R. Halmos, What does the spectral theorem say?, Amer. Math. Monthly, 70 (1963), 241-247.

[5] D. Larson, On certain reflexive operator algebras, Ph.D. thesis, University of California, Berkeley, 1976.

[6] R. L. Loebl and P. S. Muhly, Analyticity and flows in von Neumann algebras, preprint.

[7] W. E. Longstaff, Generators of nest algebras, Canad. J. Math., 26 (1974), 565-575.

[8] C. E. Rickart, General Theory of Banach Algebras, Van Nostrand, Princeton, 1960.

[9] J. R. Ringrose, On some algebras of operators, Proc. London Math. Soc., 15 (1965), 61-83.

[10] J. R. Schue, The structure of hyperreducible triangular algebras, Proc. Amer. Math. Soc., 15 (1964), 766-772.

[11] B. Sz.-Nagy, On uniformily bounded linear transformations in Hilbert space, Acta Sci. Math. (Szeged), 11 (1947), 152-157.

Recently J. Stampfli has answered Question 2 in the negative. His result appears in these Proceedings. The following elementary argument is due to J. P. Williams: If $\mathcal{B}_A \supseteq K(\mathcal{H})$ then the uniform boundedness theorem applied to the maps $\alpha_k(X) = A^k X A^{-k}$ yields $\|A^k X A^{-k}\| \le M\|X\|$. Hence for $f, g \in \mathcal{H}$, $\|A^k f\| \cdot \|A^{*-k} g\| = \|A^k f \otimes A^{*-k} g\| = \|A^k (f \otimes g) A^{-k}\| \le M\|f \otimes g\| = M\|f\| \cdot \|g\|$. Thus $\|A^k\| \cdot \|A^{-k}\| = \|A^k\| \cdot \|A^{*-k}\| \le M$ and A is similar to a multiple of unitary by the Remark after Question 2.

In addition we have been able to show that $\mathcal{B}_A \neq \{A\}'$ for $A = aI + U_+$, thus answering a question raised above.

UNIVERSITY OF KANSAS

AND

UNIVERSITY OF CINCINNATI

WEIGHTED TRANSLATION SEMIGROUPS ON $L^2[0,\infty)$

Mary Embry-Wardrop and Alan Lambert

Let $\mathcal{R}_+ = [0,\infty)$, $L^2 = L^2[0,\infty)$ (complex valued), and $\mathcal{B}(L^2)$ the bounded linear operators on L^2. If φ is a continuous function from $\mathcal{R}_+$ into $\mathbf{C}$ such that φ is never 0, we define the <u>weighted translation semigroup</u> $\{S_t\}$ with <u>symbol</u> φ by the equations

$$(S_t f)(x) = \begin{cases} 0 & ; \quad x < t \\ \dfrac{\varphi(x)}{\varphi(x-t)} f(x-t); & x \geq t. \end{cases}$$

One may easily check to see that $S_0 = I$ and $S_t S_s = S_{t+s}$. The following proposition states some of the basic properties of such a semigroup. The proofs are found in [1].

PROPOSITION. Let $\{S_t\}$ be the weighted translation semigroup with symbol φ. Then

1. $\|S_t\| = \sup_x \left| \dfrac{\varphi(x+t)}{\varphi(x)} \right|$,

2. $\{S_t\}$ is strongly continuous if and only if there are numbers M and w such that for all x and t $\left| \dfrac{\varphi(x+t)}{\varphi(x)} \right| \leq Me^{wt}$, and

3. If $\{T_t\}$ is a weighted translation semigroup with symbol ρ, then $\{S_t\}$ and $\{T_t\}$ are unitarily equivalent if and only if $|\varphi/\rho|$ is constant.

Because of (3) above we can and do assume throughout this discussion that $\varphi > 0$ on $\mathcal{R}_+$.

In many ways weighted translation semigroups appear to be continuous analogs of weighted shifts. One important difference is the existence of an infinitesimal generator. Under certain restrictions on φ we can characterize this generator quite explicitly.

PROPOSITION. If φ is in $C'(\mathcal{R}_+)$, then the infinitesimal generator, G, of

$\{S_t\}$ has domain $\mathcal{D} = \left\{ f \in L^2 : f \text{ absolutely continuous}, \ f(0) = 0, \text{ and } f' - \frac{\varphi'}{\varphi} f \in L^2 \right\}$ and $Gf = -f' + \frac{\varphi'}{\varphi} f$.

Proof. A straightforward calculation shows that for all sufficiently large λ the resolvent of G is given by

$$[(\lambda - G)^{-1} g](x) = \int_0^\infty e^{-\lambda t}[S_t g](x)dt = \frac{\varphi(x)}{e^{\lambda x}} \int_0^x \frac{e^{\lambda t}}{\varphi(t)} g(t)dt .$$

Moreover, $\mathcal{D}(G) = (\lambda - G)^{-1} L^2$. Now, if $f = (\lambda - G)^{-1} g$, then obviously f is absolutely continuous and $f(0) = 0$. Also,

$$f'(x) = \left(\frac{\varphi(x)}{e^{\lambda x}}\right)' \int_0^x \frac{e^{\lambda t}}{\varphi(t)} g(t)dt + g(x)$$

$$= \frac{\varphi'(x)}{\varphi(x)} f(x) - \lambda f(x) + g(x) ,$$

i.e. $f' - \frac{\varphi'}{\varphi} f = g - \lambda f \in L^2$, so $\mathcal{D}(G) \subseteq \mathcal{D}$. Conversely, if $f \in \mathcal{D}$, then we have for

$$g = f' - \frac{\varphi'}{\varphi} f + \lambda f, \quad \frac{e^{\lambda x}}{\varphi(x)} g(x) = \frac{d}{dx} [e^{\lambda x}/\varphi(x))f(x)] .$$

Thus

$$\int_0^x \frac{e^{\lambda t}}{\varphi(t)} g(t)dt = \frac{e^{\lambda x}}{\varphi(x)} (f(x) - f(0)) = \frac{e^{\lambda x}}{\varphi(x)} f(x) ,$$

showing $\mathcal{D} = \mathcal{D}(G)$. Finally for $f = (\lambda - G)^{-1} g$ we have

$$\begin{aligned} Gf &= G(\lambda - G)^{-1} g = -(\lambda - G)(\lambda - G)^{-1} g + \lambda(\lambda - G)^{-1} g \\ &= -g + \lambda f = -\left[f' - \frac{\varphi'}{\varphi} f + \lambda f\right] + \lambda f = -f' + \frac{\varphi'}{\varphi} f . \end{aligned}$$

We shall see that the differentiability assumption on φ is, in certain prominent special cases, satisfied. The proofs of parts (1) and (2) of the following result are to be found in [1] and [2], respectively.

THEOREM. (1) $\{S_t\}$ is hyponormal if and only if $\log \varphi$ is convex,

(2) $\{S_t\}$ is subnormal if and only if there is a number α in $\mathbb{R}$

and a probability measure ρ "on" $\mathcal{R}_+$ such that

$$\varphi^2(x) = e^{\alpha x} \int_0^\infty e^{-xt} d\rho(t).$$

Note that if $\log \varphi$ is convex, then φ is differentiable except on a countable set of points and if φ^2 is an exponential times a Laplace-Stieltjes transform as in (2), then φ^2 has an analytic extension to the halfplane $\{\mathrm{Re}\, z > 0\}$.

Part (2) of the above theorem has recently been generalized by R. Frankfurt to the case where φ is only assumed continuous on $(0,\infty)$ ([4]).

We concentrate our attention now on a rather special class of weighted translation semigroups. Our motivation for looking at this particular class was the success obtained in exhibiting unicellular weighted shifts by studying strictly cyclic shifts. For a full discussion of these concepts see [5].

<u>Definition</u>.

φ is of <u>bounded kernel type</u> if

$$k_\varphi = \sup_{x \in \mathcal{R}_+} \int_0^x \left[\frac{\varphi(x)}{\varphi(t)\varphi(x-t)}\right]^2 dt < \infty .$$

Equivalently: There is a constant c such that

$$\frac{1}{\varphi^2} * \frac{1}{\varphi^2} \le c\, \frac{1}{\varphi^2} .$$

The material presented here is a partial survey of the work done in [3].

THEOREM. Let φ be of bounded kernel type. For each f in L^2 let

$$A_f = \int_0^\infty \frac{f(t)}{\varphi(t)} S_t dt .$$

Then

(1) A_f is bounded with $\|A_f\| \le k_\varphi \|f\|$

(2) $\mathcal{A}_0 = \{A_f : f \text{ in } L^2\}$ is a commutative algebra with $A_f A_g = A_{(A_f g)}$

(3) $\mathcal{A}$ = weak closure of $\mathcal{A}_0$ is the weakly closed algebra generated by $\{S_t\}_{t=0}^\infty$.

Set $H = \left\{ \lambda \in \mathbf{C} : \dfrac{e^{\lambda t}}{\varphi(t)} \in L^2 \right\}$ and let $E = \left\{ g(t) = \dfrac{e^{\lambda t}}{\varphi(t)} : \lambda \in H \right\}$.

THEOREM. If m is a multiplicative linear functional on $\mathcal{A}$, then there is a unique g in L^2 such that (1) $m(A_f) = (f,g)$ and (2) $A_f^* g = (g,f)g$.

Conversely, if m and g satisfy (1) and (2) and $g \neq 0$, then $A^*g = (g,Ag)/\|g\|^2$ for every A in $\mathcal{A}$, and m can be extended to a multiplicative linear functional on all of $\mathcal{A}$ satisfying $m(A) = (Ag,g)/\|g\|^2$.

If m and g are as in (1) and (2) above and $g \neq 0$, it is easily shown that g is a member of E. Conversely, any g in E defines a multiplicative linear functional on $\mathcal{A}$. These observations lead to the following result.

THEOREM. For each f in L^2, $\sigma(A_f) = \{(f,g) : g \in H\} \cup \{0\}$.

It happens that there are examples of weighted translation semigroups with symbols of bounded kernel type such that H is empty; i.e., for each f, $\sigma(A_f) = \{0\}$. Such an example is exhibited later in these notes. Set $\alpha(\varphi) = \sup\{\mathrm{Re}\ \lambda : \lambda \text{ in } H\}$.

LEMMA. (1) $\alpha(\varphi) < \infty$,

(2) If $i(\varphi) = \liminf\limits_{t\to\infty} \dfrac{\varphi'(t)}{\varphi(t)}$ and $s(\varphi) = \limsup\limits_{t\to\infty} \dfrac{\varphi'(t)}{\varphi(t)}$,

then $i(\varphi) \leq \alpha(\varphi) \leq s(\varphi)$, assuming φ is in $C^1[a,\infty)$ for some $a \geq 0$.

COROLLARY. If $\lim\limits_{t\to\infty} \dfrac{\varphi'(t)}{\varphi(t)}$ exists, then this limit is $\alpha(\varphi)$.

If φ is of bounded kernel type then every densely defined linear transformation commuting with $\mathcal{A}$ is closable. To see this suppose T is such a transformation with domain $\mathcal{D}(T)$. Suppose $\{h_n\}$ is a sequence in $\mathcal{D}(T)$ such that $h_n \to 0$ and $Th_n \to f$ for some f in L^2. If $u \in \mathcal{D}(T)$ and $v \in L^2$, then $A_u v = A_v u \in \mathcal{D}(T)$. Let $g \in \mathcal{D}(T)$. Then

$$\|TA_{h_n} g\| = \|A_{h_n} Tg\| \leq k_\varphi \|h_n\|\ \|Tg\| \to 0 .$$

But $TA_{h_n} g = TA_g h_n = A_g Th_n \to A_g f$. Thus $A_g f = 0$ for every g in the dense set $\mathcal{D}(T)$. But $A_g f = A_f g$, so $A_f = 0$ and consequently $f = 0$.

Using this result it is quite easy to show that if $\alpha(\varphi) > -\infty$, then any transitive algebra containing $\mathcal{A}$ is weakly dense in $\mathcal{B}(L^2)$.

We conclude these notes with two examples of functions of bounded kernel type.

Example 1.

$\varphi(x) = x + 1$. Integration by partial fractions shows that

$$\int_0^x \left[\frac{\varphi(x)}{\varphi(t)\varphi(x-t)}\right]^2 dt = 2(x+1)^2\left[\frac{\log(x+1)}{(x+1)^3} + \frac{x}{(x+2)^2(x+1)}\right],$$

bounded on $\mathcal{R}_+$. Since $\dfrac{\varphi'(x)}{\varphi(x)} = \dfrac{1}{x+1} \to 0$ as $x \to \infty$, $\alpha(\varphi) = 0$.

Example 2.

$\varphi(x) = e^{-x^2/2}$. Note first that since $\dfrac{e^{\lambda x}}{e^{-x^2/2}}$ is never in L^2, $\alpha(\varphi) = -\infty$.

To see that φ is of bounded kernel type note that

$$\int_0^x \left[\frac{\varphi(x)}{\varphi(t)\varphi(x-t)}\right]^2 dt = e^{-x^2/2}\int_0^x e^{s^2/2}\,dx .$$

A judicious use of L'Hospital's rule shows that this is a bounded function.

REFERENCES

[1] M. R. Embry and Alan Lambert, Weighted translation semigroups, to appear in Summer 1977 Edition of Rocky Mountain Math. J.

[2] M. R. Embry and Alan Lambert, Subnormal weighted translation semigroups, J. Functional Anal., 24, No. 3, (1977), 268-275.

[3] M. R. Embry and Alan Lambert, The structure of a special class of weighted translation semigroups, to appear in Pacific J. Math.

[4] R. Frankfurt, Quasicyclic subnormal semigroups, to appear in Canad. J. Math.

[5] A. Shields, Weighted Shift Operators and Analytic Function Theory, Topics in Operator Theory (C. Pearcy, Editor), A.M.S., Providence, R.I., 1974.

UNIVERSITY OF NORTH CAROLINA AT CHARLOTTE

WEIGHTED TRANSLATION AND WEIGHTED SHIFT OPERATORS

Donald W. Hadwin and Thomas B. Hoover

The study of weighted translation operators overlaps with at least two of the three main topics of this conference, namely, composition operators and multiplication operators. Weighted translation operators were first studied by S. K. Parrott [8]. They were also studied by J. J. Bastian [1]. It is our purpose to show that each weighted translation operator is naturally associated with a family of weighted shift operators and that many problems about weighted translations can be reduced to questions about weighted shifts.

Suppose that $(X,\mathfrak{M},\mu)$ is a sigma-finite measure space. To avoid some trivial but annoying measure-theoretic pathology, we shall assume that X is a reasonably nice measure space (e.g., X is a Borel subset of the Hilbert cube and μ is a regular Borel measure). This assumption is only one of convenience; it could be completely avoided by turning to the measure algebra associated with $(X,\mathfrak{M},\mu)$.

Next, suppose that $\tau : X \to X$ is an invertible measure-preserving transformation (i.e., τ is one-to-one, onto, and $\mu(E) = \mu(\tau(E)) = \mu(\tau^{-1}(E))$ for every E in $\mathfrak{M}$). Let U_τ denote the (unitary) composition operator on $L^2(\mu)$ defined by $U_\tau f = f \circ \tau^{-1}$.

Suppose $\varphi \in L^\infty(\mu)$ and let M_φ denote the multiplication operator on $L^2(\mu)$ defined by $M_\varphi f = \varphi f$. Let $T = U_\tau M_\varphi$. An operator on $L^2(\mu)$ of the form $U_\tau M_\varphi$ is a <u>weighted translation operator</u>.

Weighted translation operators are natural generalizations of weighted shift operators. To see this, let X be the set of integers, μ be counting measure, and let $\tau(n) = n + 1$ for each n in X. Then $T = U_\tau M_\varphi$ is a bilateral weighted shift whose sequence of weights are $\{\varphi(\tau^n(0))\}$. In fact, if e_n denotes the characteristic function of $\{n\}$ for each n in X, then $\{e_n : n \in X\}$ is an orthonormal basis for $L^2(\mu)$ and $Te_n = \varphi(\tau^n(0))e_{n+1}$ for each n in X.

All weighted translation operators look pointwise like bilateral weighted shifts, i.e., if $x \in X$, then $T\chi_{\{\tau^n(x)\}} = \varphi(\tau^n(x))\chi_{\{\tau^{n+1}(x)\}}$ for each integer n. (Here

χ_E denotes the characteristic function of the set E.) To be slightly more honest we should make sure that T actually behaves pointwise like an infinite-dimensional weighted shift. Thus we shall assume that τ is <u>antiperiodic</u>; i.e., we assume that $\mu(\{x : \tau^n(x) = x\}) = 0$ for $n = 1,2,\ldots$.

For each x in X let S_x denote the bilateral weighted shift with weights $\{\varphi(\tau^n(x))\}$. It is our purpose to show that there is a very strong relationship between the operator $T = U_\tau M_\varphi$ and the shifts S_x. Clearly, we cannot hope for a relationship between T and every S_x; we can alter φ and τ on a set of measure zero without changing T.

The preceding example shows that every bilateral weighted shift can be realized (nontrivially) as one of these S_x's. However, the measure of that example is infini· What happens if the measure is finite? One answer is that almost all of the weight sequences must be "nearly" finite. More precisely, a sequence $\{a_n\}_{n=-\infty}^{\infty}$ is <u>nearly periodic</u> if there are sequences $\{n_k\}$, $\{m_j\}$ of integers such that $m_j \to -\infty$, $n_k \to +\infty$, and $|a_{i+m_j} - a_i| \to 0$, $|a_{i+n_k} - a_i| \to 0$ for each integer i. Intuitively, this means that if you are given any "block" in the sequence, then you can go arbitrarily far out (in either direction) into the sequence and find "blocks" that are arbitrarily close to the given "block". The following theorem is essentially a consequence of the recurrence theorem. We do not know if every bilateral weighted shift with a nearly periodic weight sequence can appear (nontrivially) as an S_x on a finite measure space.

THEOREM 1. If $\mu(X) < \infty$, then $\{\varphi(\tau^n(x))\}$ is nearly periodic for almost every x in X.

<u>Proof</u>. For each integer n define $\varphi_n : X \to \mathbf{C}^{(n)}$ by $\varphi_n(x) = (\varphi(x),\varphi(\tau(x)),\ldots$ $\varphi(\tau^n(x)))$. It follows that there is a set E with $\mu(E) = 0$ such that $\varphi_n(X - E)$ is contained in the essential range of φ_n for each integer n. Suppose n is a fixed integer and $\varepsilon > 0$. Partition the essential range of φ_n into finitely many Borel sets with diameter less than ε, let F be one of these sets, and let $V = \varphi_n^{-1}(F)$. It follows from the recurrence theorem [5, p.10] that, for almost every y in V, we have $\tau^k(y) \in V$ for infinitely many positive and infinitely many negative integers k. It therefore follows that, for almost every x in X,

$\|\varphi_n(x) - \varphi_n(\tau^k(x))\| < \varepsilon$ for infinitely many positive and infinitely many negative integers k. Hence $\{\varphi(\tau^n(x))\}$ is nearly periodic for almost every x in X.

One of the main connections between the operator $T = U_\tau M_\varphi$ and the shifts S_x is expressed in terms of representations of the C*-algebra $C^*(T)$ generated by 1,T.

THEOREM 2. For almost every x in X there is a representation $\pi_x : C^*(T) \to C^*(S_x)$ such that $\pi_x(1) = 1$ and $\pi_x(T) = S_x$.

The idea of the proof of Theorem 2 can be exhibited by a simple example. Suppose X is the unit circle with linear Lebesgue measure, μ is rotation by an angle α where $\alpha/2\pi$ is irrational, and suppose φ is continuous. Choose x in X. To show the existence of the representation π_x, it suffices to show that $\|p(S_x,S_x^*)\| \leq \|p(T,T^*)\|$ for every non-commutative polynomial p. Therefore it suffices to show that $\|p(S_x,S_x^*)f\| \leq \|p(T,T^*)\|$ for every non-commutative polynomial p and every unit vector f that is a finite linear combination of the orthonormal basis vectors that are shifted by S_x. It follows that we need show only that, for each positive integer n and each positive number ε, there are orthonormal vectors $\{g_k : |k| \leq n\}$ in $L^2(\mu)$ such that $\|Tg_k - \varphi(\tau^k(x))g_{k+1}\| + \|T^*g_k - g(\tau^{k-1}(x))g_{k-1}\| < \varepsilon$ for $|k| < n$. Choose a small arc J centered at x so that the arcs $\tau^k(J)$, $|k| \leq n$, are disjoint. Let $g_k = (1/\mu(J)^{1/2})\chi_{\tau^{-k}(J)}$ for $|k| \leq n$. It is clear from the continuity of φ that the desired inequalities can be obtained by choosing the length of J to be sufficiently small.

The proof in the general case replaces the continuity of φ with the consideration of the essential ranges of the φ_n's defined in the proof of Theorem 1, and with approximating the φ_n's by appropriately chosen simple functions. The analogue of choosing an arc with a lot of disjoint translates is obtained by modifying Lemma 2 in [5].

It follows from Theorem 2 that if T has some property that is preserved under representations, then almost all of the S_x's must also have that property. In order for us to be able to go in the reverse direction, we must know whether the direct sum of the π_x's is faithful. (This question could also be formulated in terms of the direct integral of the S_x's.)

THEOREM 3. The direct sum of the π_x's is a *-isomorphism.

In the case when $L^2(\mu)$ is separable even more is true. Two operators A, B are approximately equivalent (denoted by $A \sim_a B$) if there is a sequence $\{U_n\}$ of unitary operators such that $U_n^* A U_n - B$ is compact for $n = 1,2,\ldots,$ and $\|U_n^* A U_n - B\| \to 0$ as $n \to \infty$ (i.e., if A is unitarily equivalent to arbitrarily small compact perturbations of B). The following theorem is based on the preceding theorem and the beautiful results of D. Voiculescu [10] on approximate equivalence.

THEOREM 4. If $L^2(\mu)$ is separable, then there are points $x_1, x_2, \ldots$ in X such that $T \sim_a S_{x_1} \oplus S_{x_2} \oplus \cdots$.

In the preceding theorem T is also approximately equivalent to the direct integral of the S_x's.

There is an interesting consequence of Theorem 4. If the weights in a weighted shift are replaced by their absolute values, then the new shift is unitarily equivalent to the original one. One might guess that $U_\tau M_\varphi$ is always unitarily equivalent to $U_\tau M_{|\varphi|}$. However, S. K. Parrott [6, Example 2.1] has shown that this is false. The preceding theorem implies that if $L^2(\mu)$ is separable, then $U_\tau M_\varphi$ is always approximately equivalent to $U_\tau M_{|\varphi|}$.

So far we have only considered what happens when we put restrictions on μ. What happens if we put restrictions on τ? We have already assumed that τ is invertible and measure-preserving, so there aren't many restrictions left. One of the few that come to mind is ergodicity. Ergodicity is a sort of irreducibility condition. The transformation τ is ergodic if it is impossible to write X as a disjoint union of two sets X_1, X_2 of positive measure such that $\tau(X_1) \subseteq X_1$ and $\tau(X_2) \subseteq X_2$. An example of an ergodic transformation is the irrational rotation of the circle considered earlier.

If τ is ergodic, then almost all of the S_x's are approximately equivalent.

THEOREM 5. If τ is ergodic, then there is an x_0 in X such that $S_x \sim_a S_{x_0}$ for almost all x in X.

If τ is ergodic, then it follows from Theorems 4 and 5 that $T \sim_a S_x \oplus S_x \oplus \cdots$

for almost every x in X. However, it follows from Voiculescu's results on approximate equivalence that more is true.

THEOREM 6. If τ is ergodic, then $T \sim_a S_x$ for almost every x in X.

Note that none of the preceding results depend upon φ.

Using the preceding results, we have determined which weighted translation operators are similar to a normal operator. The proof uses the fact [4, Theorem 6.3] that if an operator A is similar to a normal operator and π is a representation of $C^*(A)$, then $\pi(A)$ is similar to a normal operator. It is not difficult to show that a bilateral weighted shift S is similar to a normal operator if and only if it is similar to $r(S)$ times the (unweighted) bilateral shift; here $r(S)$ denotes the spectral radius of S. The following theorem extends a theorem of S. K Parrott [8, Theorem 3.7] that characterizes the weighted translation operators that are similar to unitary operators (see also [9]). Note that if an operator A is similar to a normal operator, then A and A^2 have the same kernel; it follows that if a weighted translation operator T is similar to a normal operator, then the kernel of T reduces T. Thus the restriction that $\ker T = 0$ in the following theorem is not as severe as it sounds.

THEOREM 7. Suppose $\ker T = 0$. Let $c(x) = r(S_x)$ for every x in X. The following are equivalent.

(1) T is similar to a normal operator,

(2) there is an invertible h in $L^\infty(\mu)$ such that $M_h^{-1} T M_h = U_\tau M_{c\cdot \operatorname{sgn}(\varphi)}$,

(3) there are positive numbers a,b such that

$$a \leq |\varphi(x)\varphi(\tau(x)) \cdots \varphi(\tau^n(x))|/c(x)^{n+1} \leq b$$

for almost every x in X and for $n = 1,2,\ldots,$

(4) there is an invertible g in $L^\infty(\mu)$ such that $\varphi(x)/c(x) = g(\tau(x))/g(x)$,

(5) the direct sum of the $\pi_x(T)$'s is similar to a normal operator.

Many of the preceding theorems are true in a more general setting. Suppose H is a separable Hilbert space, and let $L^2(\mu;H)$ denote the set of all Borel measurable functions $f: X \to H$ such that $\int_X \|f\|^2 d\mu < \infty$. Then $L^2(\mu;H)$ is a Hilbert space with inner product $(f,g) = \int_X (f(x),g(x))d\mu(x)$. (Actually, we must identify two functions if they disagree only on a set of measure zero.) If we define U_τ on $L^2(\mu;H)$ by $U_\tau f = f \circ \tau$, then U_τ is a unitary operator. Next suppose $\varphi : X \to B(H)$ is a weakly Borel measurable, essentially (norm) bounded function and define M_φ on $L^2(\mu;H)$ by $(M_\varphi f)(x) = \varphi(x)f(x)$ for every x in X. For each x in X, let S_x be the bilateral weighted shift with operator weights $\{\varphi(\tau^n(x))\}$. If $T = U_\tau M_\varphi$, then analogues of many of the preceding theorems are true. (However, "nearly periodic" must be defined with respect to the *-strong operator topology.)

Analogues also hold if the condition that τ be measure-preserving is replaced by the condition that μ and $\mu \circ \tau^{-1}$ are mutually absolutely continuous and $d\mu/d\mu \circ \tau^{-1}$, $d\mu \circ \tau^{-1}/d\mu$ are essentially bounded. In this case the values of $d\mu/d\mu \circ \tau^{-1}$ must appear in the weights of the S_x's.

Using the preceding generalizations we have been able to answer some important questions about the C^*-algebra generated by a weighted translation operator; in fact, we can infer many facts concerning the C^*-algebra generated by a centered operator. An operator A is a centered operator [6] if the operators $\ldots, A^2(A^*)^2, AA^*, A^*A, (A^*)^2A^2, \ldots$ commute with each other. We have determined which centered operators generate GCR C^*-algebras as well as the irreducible representations of such algebras. In particular, an irreducible centered operator (on an infinite-dimensional Hilbert space) that generates a GCR C^*-algebra must be a weighted bilateral shift or a weighted (forward or backward) unilateral shift. Note that an operator generates a GCR C^*-algebra if and only if it is smooth in the sense of J. Ernest [2], [3]. These latter results are related to the work of O'Donovan [7].

Complete proofs of the results of this paper will appear elsewhere.

REFERENCES

[1] J. J. Bastian, Decompositions of weighted translation operators, Ph.D. Thesis, Indiana Univ., 1973.

[2] J. Ernest, Charting the operator terrain, Memoirs of the Amer. Math. Soc., No. 171, (1976).

[3] J. Ernest, Concrete representations and the von Neumann type classification of operators, these proceedings.

[4] D. W. Hadwin, An asymptotic double commutant theorem for C^*-algebras, Trans. Amer. Math. Soc., to appear, 1978.

[5] P. R. Halmos, Lectures on ergodic theory, Chelsea, New York, 1956.

[6] B. B. Morrel and P. S. Muhly, Centered operators, Stud. Math., 51 (1974), 251-263.

[7] D. O'Donovan, Weighted shifts and covariance algebras, Trans. Amer. Math. Soc., 208 (1975), 1-25.

[8] S. K. Parrott, Weighted translation operators, Ph.D. Thesis, Univ. of Michigan, 1965.

[9] K. Petersen, The spectrum and commutant of a certain weighted translation operator, Math. Scand., 37 (1975), 297-306.

[10] D. Voiculescu, A non-commutative Weyl-von Neumann theorem, Rev. Rom. Math. Pures et Appl., 21 (1976), 97-113.

UNIVERSITY OF NEW HAMPSHIRE

UNIVERSITY OF HAWAII

AN OPERATOR NOT A SHIFT, INTEGRAL, NOR MULTIPLICATION

Bernard N. Harvey

Dr. Paul Halmos has conjectured that the only concrete models we have for representing abstract linear operators in Hilbert space are unitarily equivalent to either a shift, a multiplication, or an integration operator. The following is a concrete linear operator which shows this conjecture false.

Let $L^2 = L^2(1,2)$ be the space of all Lebesgue measurable functions which are square integrable with the usual inner product and let $H = L^2 \dotplus L^2$, the direct sum of L^2 with itself. Consider the operator defined in H by

$$T(t) = \begin{pmatrix} t & 1 \\ 0 & t \end{pmatrix}.$$

T is not normal since it has a nilpotent part and therefore cannot be unitarily equivalent to a multiplication. It is not an isometry and hence cannot be unitarily equivalent to a shift. The fact that it is not unitarily equivalent to an integral operator with Carleman kernel is due to the following two results. A theorem due to von Neumann states that a self-adjoint operator is unitarily equivalent to an integral operator with Carleman kernel if and only if 0 belongs to the set of limit points of its spectrum. This set includes the continuous spectrum, limit points of the point spectrum, and characteristic values of infinite multiplicity. A theorem stated by Dr. Halmos at this symposium is: A (non-normal) linear transformation is unitarily equivalent to a Carleman integral operator if and only if the Hermitian part of its polar factorization is unitarily equivalent to a Hermitian Carleman integral operator. A calculation shows that TT^* is given by

$$TT^* = \begin{pmatrix} t^2+1 & t \\ t & t^2 \end{pmatrix}.$$

This operator has no point spectrum and so the set of limit points of its spectrum is merely its continuous spectrum. Calculating the resolvent of TT^*, we see that the continuous spectrum is all z such that $z = \frac{1}{2}\left(2t^2 + 1 \pm \sqrt{4t^2+1}\right)$ for t between 1 and 2. If $f(t) = 2t^2 + 1 - \sqrt{4t^2+1}$ then $f'(t) \geq 0$ and equals 0 only when $t = 0$. Thus f is strictly increasing from 1 to 2 and $f(t) \geq f(1) = 3 - \sqrt{5} > 0$. Therefore 0 is not in the set of limit points of the spectrum of TT^* and the same is true for the square root of TT^*. By von Neumann's result $\sqrt{TT^*}$ is not unitarily equivalent to a Carleman integral operator and therefore by Halmos' result neither is T.

As the three types of operators mentioned by Dr. Halmos have an abstract characterization I have attempted to do the same for T. At present I have such a characterization but it is very cumbersome and not worth reproducing here. However the point is that it can be done.

CALIFORNIA STATE UNIVERSITY LONG BEACH

STRICTLY CYCLIC OPERATOR ALGEBRAS AND

APPROXIMATION OF OPERATORS

D. A. Herrero

Let $\mathcal{L}(\mathcal{H})$ denote the algebra of all bounded linear operators acting on a complex separable infinite dimensional Hilbert space $\mathcal{H}$. C. Foiaş, C. Pearcy and D. Voiculescu [3] proved that the subset

$$(BQT)_{qs} = \{A \in \mathcal{L}(\mathcal{H}) : A \text{ is quasisimilar to some biquasitriangular operator}\}$$

is (norm) dense in $\mathcal{L}(\mathcal{H})$ and asked whether $(BQT)_{qs}$ is actually equal to $\mathcal{L}(\mathcal{H})$.

The answer is negative. Furthermore, suitable modifications of the Apostol-Morrel models [2] provide the following result

THEOREM A. The complement of $(BQT)_{qs}$ is also dense in $\mathcal{L}(\mathcal{H})$.

The main ingredient of the proof is the construction of large family of operators with very particular properties. Let Ω be a nonempty bounded connected open subset of the plane such $\partial\Omega$ consists of finitely many pairwise disjoint regular analytic Jordan curves and let $\Lambda = \{\lambda_1, \lambda_2, \ldots, \lambda_m\}$ be a finite subset of $\mathbb{C}\setminus\Omega^-$ having exactly one point in each component of this set. Let $\varepsilon > 0$ be small enough so that $\Lambda \cap (\Omega^- + \varepsilon t) = \emptyset$ for $0 \leq t \leq 1$ and define $\Gamma = \{(z,t) \in \mathbb{C}\times(0,1) : z - \varepsilon t \in \partial\Omega\}$. (The "leaning tower".) Let $W^{2,2}(\Gamma)$ be the Sobolev space of all distributions on the analytic manifold Γ whose partial derivatives up to order 2 belong to $L^2(\Gamma, dm)$ (dm is the "area measure" induced by 3-dimensional Lebesgue measure).

$W^{2,2}(\Gamma)$ can be identified with a Banach algebra (under an equivalent norm) of continuous functions on Γ^- under pointwise operations [1].

This algebra contains a subalgebra

$A^{2,2}(\Gamma) = \text{closure } \{f(z,t) = \sum_{k=0}^{n} t^k f_k(z) : n = 1,2,\ldots\}$ consisting of all " analytic elements" of $W^{2,2}$. (Here f_k denotes an arbitrary rational function of z with poles in Λ.)

Let $T \in \mathfrak{L}(W^{2,2})$ be the operator "multiplication by z" ($Tf(z,t) = zf(z,t)$). Clearly, $A^{2,2}(\Gamma)$ is invariant under T and (1) $\mathfrak{a}'(T)$ (= the commutant of T) $\supset \{M_g : g \in W^{2,2}\}$ (where M_g = "multiplication by g") and (2). If $L = T \mid A^{2,2}(\Gamma)$, then $\mathfrak{a}'(L) \supset \{M_g : g \in A^{2,2}(\Gamma)\}$. Let $\sigma(T)$, $E_\ell(T)$ and $E_r(T)$ denote the spectrum, the left essential spectrum and the right essential spectrum of T, respectively.

THEOREM B. With the above notation:

(i) $\mathfrak{M}[W^{2,2}(\Gamma)]$ (= Gelfand spectrum) $= \overline{\Gamma}$.

(ii) $\mathfrak{M}[A^{2,2}(\Gamma)] = \{(z,t) \in \mathbb{C} \times [0,1] : z - \varepsilon t \in \Omega^-\}$.

(iii) $\sigma(T) = E_\ell(T) = E_r(T) = E_\ell(L) = \{z : (z,t) \in \Gamma^-\}$.

(iv) $\sigma(L) = E_r(L) = \{z : (z,t) \in \mathfrak{M}[A^{2,2}(\Gamma)]\}$.

(v) $\text{Ker}(\lambda - L) = \{0\}$ and $\dim \text{Ker}(\lambda - L)^* = \infty$ for every $\lambda \in \sigma(L) \setminus E_\ell(L)$.

(vi) If $e(z,t) \equiv 1$, then $A^{2,2}(\Gamma) = \mathfrak{a}''(L)e = \{Ae : A \in \mathfrak{a}''(L)\}$, where $\mathfrak{a}'(L)$ and $\mathfrak{a}''(L)$ denote the commutant and the double commutant of $L \in \mathfrak{L}(A^{2,2}(\Gamma))$, resp. This implies $\mathfrak{a}''(L) = \mathfrak{a}'(L)$ is a maximal abelian strictly cyclic subalgebra of $\mathfrak{L}(A^{2,2}(\Gamma))$ with strictly cyclic vector e.

(vii) If L' is quasisimilar to L, then L' is actually similar to L.

Proof. (i)-(v) are standard facts.

(vi) Let $A \in \mathfrak{a}'(L)$ and let $\eta, \tau \in [0,1]$, $\eta \neq \tau$.

Assuming that $|\eta - \tau| > 8\delta > 0$, let $h_\eta(z,t) = H_\eta(t) \in A^{2,2}(\Gamma)$ be defined by

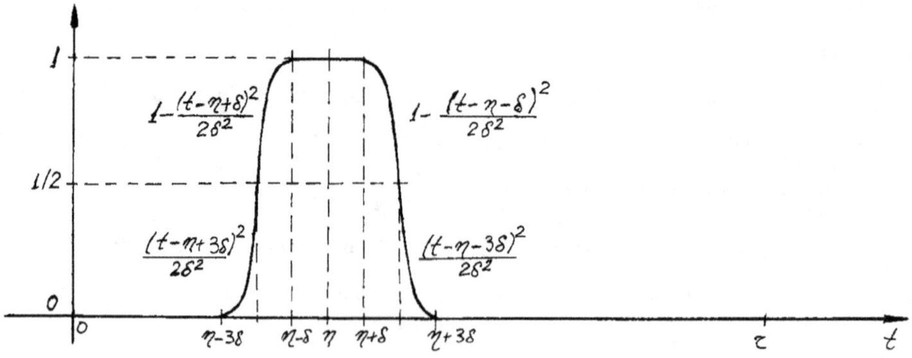

and let M_η be the multiplication by h_η. Let h_τ, H_τ and M_τ be similarly defined. Clearly, $M_\eta A M_\tau$ belongs to $\mathfrak{a}'(L)$.

Define $\psi : [0,1] \to [\tau - 4\delta, \tau + 4\delta] \cap [0,1]$ to be an arbitrary C^∞ function, bijective, such that

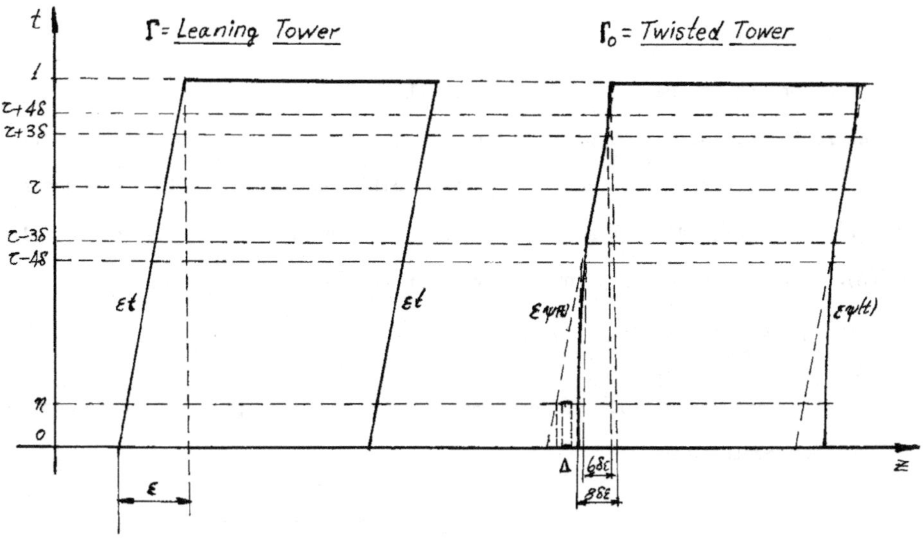

$\psi(t) = t$ in $[\tau - 3\delta, \tau + 3\delta] \cap [0,1]$ and $\min\{\psi'(t) : t \leftarrow [0,1]\} > 0$. Define $S : W^{2,2}(\Gamma_0) \to W^{2,2}(\Gamma)$ as the isomorphism mapping every function "in the same level", where $\Gamma_0 = \{(z,t) : z - \varepsilon\psi(t) \in \partial\Omega,\ 0 < t < 1\}$. (Roughly: S "untwists" Γ_0 onto Γ.)

Finally, let Δ be a small closed disc contained in the "η-level" of $\mathfrak{M}(A^{2,2}(\Gamma))$ which does not intersect the projection of Γ_0 onto $\mathbf{C}$. If $R : A^{2,2}(\Gamma) \to H^2(\Delta)$ is defined by $Rf(z,t) = f(z,\eta) \mid \Delta$, then R has dense range and, since Ω is <u>connected</u>, $f(z,\eta)$ is uniquely determined by $f(z,\eta) \mid \Delta$.

Since $M_\eta A M_\tau \in \mathfrak{a}'(L)$, we have

$$L(M_\eta A M_\tau) - (M_\eta A M_\tau)L = 0 ,$$

and therefore

$$0 = RLM_\eta AM_\tau S - RM_\eta AM_\tau LS = L_\Delta(RM_\eta AM_\tau S) - (RM_\eta AM_\tau S)\, L_0 ,$$

where $L_\Delta(L_0)$ is the "multiplication by z" on $H^2(\Delta)$ (on $A^{2,2}(\Gamma_0)$, resp.).

Since $\sigma(L_\Delta) \cap \sigma(L_0) = \Delta \cap (\mathrm{Proj}_{\mathbf{C}}(\Gamma_0)) = \emptyset$, we conclude from Rosenblum's Corollary [7, Corollary 0.13] that $RM_\eta AM_\tau S = 0$. Since S is an isomorphism, we obtain $RM_\eta AM_\tau = 0$, and therefore

$$(M_\eta AM_\tau f)\ (z,\eta) = (AM_\tau f)\ (z,\eta) = 0$$

for all $f \in A^{2,2}(\Gamma)$.

A second approximation argument shows that $A = M_g$, where $g = Ae$. (Consider A restricted to suitable approximations in $A^{2,2}(\Gamma)$ of a function of the form $\sum_{k=0}^n t^k f_k(z)$ by functions that are constant with respect to t in a neighborhood of τ.) Therefore $\mathcal{a}'(L) = \{M_g : g \in A^{2,2}(\Gamma)\}$ is abelian; whence we obtain $\mathcal{a}'(L) = \mathcal{a}''(L)$. The remaining statement is obvious now.

(vii) This follows from (vi).

Similar arguments produce operators L with the same properties, except that (v) must be replaced by

$(v)_n$ $\mathrm{Ker}(\lambda - L) = \{0\}$ and $\dim \mathrm{Ker}(\lambda - L)^* = n > 0$ for every $\lambda \in \sigma(L) \setminus E(L)_\varepsilon$,

where $E(L)_\varepsilon = \{\lambda : \mathrm{dist}[\lambda, E(L)] \le \varepsilon\}$.

By using these results, several known results about closures of similarity orbits [4], [5] and the constructions of [2], it is possible to prove

THEOREM C. If $\mathcal{H}$ is a complex, separable, infinite dimensional Hilbert space, then $\{T \in \mathcal{L}(\mathcal{H}) : T$ is similar to $A \oplus B$, $\mathcal{a}''(A)$ and $\mathcal{a}''(B^*)$ are strictly cyclic, $\sigma(A) \cap \sigma(B) = \emptyset$, $\lambda_A - A$ and $\lambda_B - B^*$ are semi-Fredholm operators of index $-\infty$ for suitably chosen $\lambda_A, \lambda_B \in \mathbf{C}$, and every T' quasisimilar to T is actually similar to $T\}$ is dense in $\mathcal{L}(\mathcal{H})$.

Since every T in the dense set of Theorem C lies in the complement of $(BQT)_{qs}$, it readily follows that Theorem A is a corollary of the above result. The same arguments apply to many other cases. Complete proofs will appear in [6].

REFERENCES

[1] R. A. Adams, Sobolev Spaces, Academic Press, New York-San Francisco-London,

[2] C. Apostol and B. Morrel, On approximation of operators by simple models, Indiana University Math. J., 26 (1976), 427-442.

[3] C. Foiaş, C. Pearcy and D. Voiculescu, Biquasitriangular operators and quasi-similarity, (to appear).

[4] D. A. Herrero, Closure of similarity orbits of Hilbert space operators, II: Normal operators, J. London Math. Soc., (2) 13 (1976), 299-316.

[5] D. A. Herrero, Closure of similarity orbits of Hilbert space operators, III, Math. Ann., (to appear).

[6] D. A. Herrero, Quasisimilar operators with different spectra, Acta Sci. Math. (Szeged), (to appear).

[7] H. Radjavi and P. Rosenthal, Invariant subspaces, Ergebnisse der Mathematik und ihrer Grenzgebiete, B. 77, Springer-Verlag, New York-Heidelberg-Berlin, 1973.

INSTITUTO VENEZOLANO DE INVESTIGACIONES CIENTIFICAS
Departamento de Mathemáticas
A.P. 1827, Caracas 101, Venezuela

ON SINGULAR SELF-ADJOINT STURM-LIOUVILLE OPERATORS

Gerhard K. Kalisch

The interest that singular self-adjoint Sturm-Liouville operators present from the standpoint of operator theory in Hilbert space H consists in their universality: all cyclic self-adjoint operators (bounded or not) are unitarily equivalent to restrictions of singular self-adjoint Sturm-Liouville operators and conversely. To put the facts in their proper setting and perspective, we present a number of known results as well as some new ones whose detailed exposition will be published elsewhere.

We assume an acquaintance with symmetric operator theory [3]. We also assume an acquaintance with elementary singular Sturm-Liouville theory on $(0,\infty)$ [2,5]. To establish our notation and definitions, let $D_0 = D_0(S_0)$ be the domain of definition of the symmetric operator S_0 and $D_0^* = D_0^*(S_0^*)$ be the domain of definition of its adjoint S_0^*; we write $\text{ind}(S_0) = (j,k)$ for the defect index of S_0 with j and k non-negative integers or infinity. Our operators in the sequel will have defect index $(1,1)$ and will have no self-adjoint parts; i.e., non-trivial self-adjoint restrictions to reducing subspaces. The self-adjoint extensions $S(a)$ of S_0 are defined on domains $D(a)$ with $D_0 \subset D(a) \subset D_0^*$; $a \in A$ is a suitable parametrization of these self-adjoint extensions.

Our Sturm-Liouville operators, which are defined on $L_2(0,\infty) = L_2$ with Lebesgue measure, are induced by the action of a differential expression given by $Xf = -f'' + qf$ for f in a suitable dense linear subset of L_2 where the real-valued function q is locally L_1. We define the symmetric operator L_0 induced by X on the domain $D_0 = \{f \in L_2;\ f \ \&\ f' \text{ abs. cont. } Xf \in L_2;\ f(0) = f'(0) = 0\}$. The operator L_0^* is induced by X also; its domain of definition is $D_0^* = \{f \in L_2;\ f \ \&\ f' \text{ abs. cont.}; \ Xf \in L_2\}$. The defect index of L_0 is $(1,1)$ or $(2,2)$, Weyl's limit point and limit circle case, respectively. Our interest centers on the first case only; the self-adjoint extensions of L_0 have only discrete spectrum in the second case. See [2,5] for details.

Thus let L_0 have defect index $(1,1)$; its self-adjoint extensions $L = L(q,\alpha)$

are determined by the subset $D(q,\alpha)$ of D_0^* determined by the boundary condition at 0:

$$\sin \alpha f - \cos \alpha f' = 0 \quad \text{with} \quad \alpha \in [0, a\pi).$$

The spectral theorem for $L(q,\alpha)$ then takes this form: Let $\varphi(s,z) = \varphi_\alpha$ and similarly $\psi(s,z) = \psi_\alpha$ be solutions of $\mathcal{X}f = zf$ $(z \in \mathbb{C})$ with boundary conditions at 0 given by $\varphi_\alpha = \sin \alpha$, $\varphi'_\alpha = -\cos \alpha$; $\psi_\alpha = \cos \alpha$, $\psi'_\alpha = \sin \alpha$. Then there exists a unique Borel measure $\rho = \rho\ (q,\alpha)$ on $\mathbb{R}$ given by a nondecreasing function also called ρ, and an isometry $\mathfrak{F} = \mathfrak{F}(q,\alpha)$ of L_2 onto $L_2(\rho)$ given by

$$(\mathfrak{F}f)(\lambda) = \int_0^\infty \psi(s,\lambda)\, f(s)\, ds, \quad (\mathfrak{F}^{-1}F)(s) = \int_{-\infty}^\infty \psi(s,\lambda)\, F(\lambda)\, d\rho(\lambda).$$

If the self-adjoint operator M on $L_2(\rho)$ is defined by $(MF)(\lambda) = \lambda F(\lambda)$ with domain of definition $D(M) = \{F \in L_2(\rho);\ \lambda F \in L_2(\rho)\}$, then:

$$\mathfrak{F}D(q,\alpha) = D(M), \quad \mathfrak{F}^{-1}(D(M)) = D(q,\alpha),$$

$$\mathfrak{F}L(q,\alpha) = M\mathfrak{F} \quad \text{and} \quad \mathfrak{F}^{-1}M = L(q,\alpha)\mathfrak{F}^{-1}.$$

Furthermore, while φ and ψ are not in L_2, there exists a function $m = m(z) = m(z;q,\alpha)$ with $\varphi + m\,\psi \in L_2$ and $m(z) = \int_{-\infty}^\infty (1/(\lambda - z))d\rho(\lambda) - \tan \alpha$ (for the cases $\tan \alpha = \infty$, see [2,5]). The function m is also related to φ and ψ by the formula $m(z) = -\lim_{a\to\infty} (\varphi(a,z)/\psi(a,z))$. Finally the measure ρ, called the <u>spectral function</u> or <u>spectral measure</u> of $L(q,\alpha)$, is expressible in terms of m by means of the formula

$$\rho(\lambda) = (1/\pi) \lim_{y\downarrow 0} \int_0^\lambda \operatorname{Im} m(x + iy)\, dx.$$

As an example, consider the case $q = \alpha = 0$ and call the corresponding spectral measure ρ_0. This measure is zero on the negative reals and equals $(2/\pi)\sqrt{\lambda}$ on the positive reals; the corresponding function ψ equals $\cos(\sqrt{\lambda}\ s)$ and the corresponding isometry $\mathfrak{F}_0$ is then essentially the cosine transform.

The fact that the Sturm-Liouville operators under consideration are unitarily equivalent to operators "multiplication by the independent variable" shows that only cyclic operators are representable in this manner. The question then naturally arises, which cyclic self-adjoint operators are unitarily equivalent to these differential operators. This is the "inverse Sturm-Liouville" problem. The history of this problem is well discussed in [3,7]. We base our exposition on the Levitan-Gasymov [7] paper which contains the best formulation of the results and proofs that we know.

The universality of the singular self-adjoint Sturm-Liouville operators comes from the consequence of the basic Gelfand-Levitan theorem which asserts that all cyclic self-adjoint operators (bounded or not) are unitarily equivalent to parts of Sturm-Liouville operators. The analogous problem for the general case — even multiplicity two — is completely open.

INVERSE STURM-LIOUVILLE THEOREM. A necessary and sufficient condition that a Borel measure ρ on R be the spectral measure of the singular self-adjoint Sturm-Liouville operator $L(q,\alpha)$ is that

(1) If $f \in L_2$ has compact support, then the vanishing of its cosine transform $\int_0^\infty f(s)\cos(\sqrt{\lambda}\, s)\, ds$ on the support of ρ implies that $f = 0$ a.e., and

(2) the family of functions $\varphi_N(s) = \int_0^N \cos(\sqrt{\lambda}\, s)\, d(\rho_0 - \rho)(\lambda)$ converges boundedly to a function $\varphi(s)$ which has a locally L_1 derivative.

Several comments are in order.

(1) Corresponding ρ's and $L(q,\alpha)$'s determine each other uniquely.

(2) The sufficiency part of the theorem is not just an existence and uniqueness statement; the proof provides an actual construction of β and α when ρ is given. For subsequent purposes of the present account, we quote the formulas that exhibit this connection.

(a) Condition (2) of the theorem implies the existence of the function $F(x,y) = \int_{-\infty}^{\infty} \cos(\sqrt{\lambda}\, x)\cos(\sqrt{\lambda}\, y)\, d(\rho_0 - \rho)(\lambda)$.

(b) A function $K(x,y)$ is determined uniquely by F via the <u>Gelfand-Levitan equation</u>

$$F(x,y) - K(x,y) + \int_0^x K(x,s)F(s,y)\, ds = 0 .$$

The function K is as smooth as F.

(c) The function q and the real number α depend on K as follows:

$$K(x,x) = \tan \alpha + \frac{1}{2} \int_0^x q(s)\, ds \,;$$

when $\tan \alpha = \infty$, these formulas are modified.

(3) $q \in C^{(m-1)}$ and $q^{(m)} \in \text{loc } L_1$ if and only if $\varphi \in C^{(m)}$ and $\varphi^{(m+1)} \in \text{loc } L_1$.

(4) Condition (1) is not needed unless ρ has discrete support. In that case, the condition is equivalent to $\limsup_{N\to\infty} [(\text{number of positive eigenvalues} < \,)/\sqrt{N}\,] = \infty$ (see [6,7]).

There are various sufficient conditions for a cyclic self-adjoint operator to be unitarily equivalent to a Sturm-Liouville operator.

THEOREM S. (1) The cyclic self-adjoint operator H is unitarily equivalent to a Sturm-Liouville operator if and only if its positive part (i.e., its part with non-negative spectrum) is; if H has its spectrum bounded from above, it cannot be equivalent to a Sturm-Liouville operator.

(2) The cyclic self-adjoint operator H is unitarily equivalent to a Sturm-Liouville operator if:

(i) it has a part that is unitarily equivalent to one; or

(ii) its continuous spectrum contains an infinite positive interval; or

(iii) the positive part of its discrete spectrum $\{\lambda_n\}$ satisfies condition

(C) its spectral function ρ can be so normalized that $\rho_0 \geq \rho$ for $\lambda \geq a$ and $\rho_0 - \rho \in L_1(a,\infty)$; or

(iv) the complement of its spectrum, $\cup I_j$, with $I_j = (\lambda_j, \mu_j)$, is such that $\{\lambda_n\}$ satisfies condition (C) in (iii).

UNIVERSALITY THEOREM. Every cyclic self-adjoint operator is unitarily equivalent to a part of a Sturm-Liouville operator. If it is bounded, the part of the Sturm-Liouville operator to which it is equivalent has for its domain of definition a Hilbert space of entire functions.

The last part of the preceding theorem is a consequence of the Paley-Wiener Theorem.

Another question that arises naturally is, given a symmetric operator with defect index (j,j), what are all its self-adjoint extensions? Or, given a family of self-adjoints, what are necessary and sufficient conditions that it should be the family of all self-adjoint extensions of some (several?) symmetric operators? What is a complete set of unitary invariants of symmetric operators, say, with defect index (j,j), or even $(1,1)$? In this last case, if the symmetric operator has no non-trivial self-adjoint part, there are some partial answers.

WEYL'S THEOREM ON THE CONTINUOUS SPECTRUM. Any two self-adjoint extensions of a symmetric operator with defect index $(1,1)$ and no self-adjoint part have the same continuous spectrum.

This has been refined by Aronszajn and Donohue [1,4].

THEOREM A-D. The absolutely continuous parts of two self-adjoint extensions as in Weyl's Theorem above are unitarily equivalent; the continuous singular parts are mutually singular.

A counterpart of Weyl's theorem deals with pure point spectra; for examples of Sturm-Liouville operators with pure point spectra see [8]. The first theorem dealing with the point spectra of two self-adjoint extensions of the same (regular) Sturm-Liouville operator is due to Borg; see [7]. In effect they determine the symmetric operator and the two extensions. This was extended to the general case with pure point spectrum by Donoghue and by the present author to the general case of singular spectra.

THEOREM ON TWO SPECTRA. The unitary invariants of two singular mutually singular self-adjoint extensions H_1 and H_2 of the symmetric operator S with defect index $(1,1)$ and no self-adjoint part determine S, H_1, and H_2 uniquely.

To determine when two such unitary invariants can occur and to construct the operators in question — say, to construct q, α_1, α_2 in the case of Sturm-Liouville operators — is much deeper. The answer is completely known in the regular Sturm-

Liouville case and partially in the singular case (only in connection with pure point spectra) [7].

The relations between $L(q,\alpha)$, ρ, and m which were discussed at the beginning of this paper, and the formulas quoted after the statement of the Inverse Sturm-Liouville Theorem furnish the basis for our last theorem — we use our earlier notation and conventions.

APPROXIMATION THEOREM. The weak* convergence of the measures ρ_n to the measure ρ_0 is equivalent to the convergence (uniformly on compacts) of the corresponding analytic functions $m_n(z)$ to $m_0(z)$ and to the convergence of the pair (q_n,α_n) to (q_0,α_0) where the convergence of the q_n is uniform on compacts and that of the α_n is ordinary convergence of real numbers.

A type of convergence of the corresponding operators $L(q_n,\alpha_n)$ to $L(q_0,\alpha_0)$ will be introduced in another paper in a general context.

REFERENCES

[1] N. Aronszajn, On a problem of Weyl in the theory of singular Sturm-Liouville equation, Amer. J. Math., 79 (1957(, 597-610.

[2] E. A. Coddington and Norman Levinson, Theory of Ordinary Differential Equations, McGraw-Hill, New York, 1955.

[3] Nelson Dunford and J. T. Schwartz, Linear Operators I & II 1963.

[4] W. F. Donoghue, On the perturbation of spectra, Comm. on Pure & Appl. Math., 18 (1965), 559-579.

[5] Einar Hille, Lectures on Ordinary Differential Equations, Addison-Wesley, Reading, 1969.

[6] B. Ja. Levin, Distribution of Zeros of Entire Functions, Translations of Math. Monographs, 5, Amer. Math. Soc., Rhode Island, 1964.

[7] B. M. Levitan and M. G. Gasymov, Determination of a differential equation by two of its spectra, Usp. Mat. Nauk, (2), 19, 2, (1964), 3-63 = Math. Surveys, 19, 2, (1964), 1-63.

[8] E. C. Titchmarsh, Eigenfunction Expansions I, Second Edn., Oxford Univ. Press, Oxford, 1962.

UNIVERSITY OF CALIFORNIA AT IRVINE

EXTENSIONS OF COMMUTING SUBNORMAL OPERATORS*

Arthur Lubin

An operator T on a Hilbert space H is called subnormal if there exists a normal operator N defined on some larger space K containing H such that the restriction $N|H = T$. The concept of subnormality was introduced by Halmos [3], who showed that a subnormal operator has a minimal normal extension (m.n.e.) unique up to unitary equivalence, and an intrinsic necessary and sufficient condition for an operator to be subnormal was given by Halmos and Bram [2]. Two commuting subnormals S and T on H are said to have commuting normal extensions (c.n.e.) if there exist commuting normal operators M and N defined on some $K \supset H$ with $M|H = S$ and $N|H = T$. The positivity condition of Halmos and Bram was extended by Ito [6] to give a necessary and sufficient condition for c.n.e., but it was not clear whether this condition was automatically satisfied by every pair of commuting subnormals. Recently, however, examples of commuting subnormals without c.n.e. have been found independently by M. B. Abrahamse and the author [1,10].

Abrahamse's example was related to a study of when elements in the commutant of a subnormal lift to the commutant of the m.n.e. Relevant to this is the following:

BASIC LEMMA [1]. Two commuting subnormals S and T have c.n.e. if and only if there is a subnormal extension S_1 of S which commutes with the m.n.e. of T.

In this context, one would expect the problem of determining whether an extension S_1 is subnormal to be difficult, however, in the example given by Abrahamse S does not have any extension commuting with the m.n.e. of T and a fortiori no subnormal one. This motivates the following:

<u>Question</u> [1]. If S and T are commuting subnormals and if S extends to an operator S_1 which commutes with the m.n.e. of T, must S_1 be subnormal? (i.e., Does there exist c.n.e.?)

Our example below answers the question in the negative.

---

*Research supported by NSF Grant MCS76-06516.

For each multiindex $J = (j,k)$, let $\beta(J) = (j!k!)$ and let $H = H^2(\beta) = \{\Sigma f_J z^i w^k : \Sigma |f_J|^2 \beta(J)^2 < \infty\}$.

We define S and T on H by

$$S f(z,w) = z f(z,w) \quad \text{and} \quad T f(z,w) = w f(z,w).$$

These maps, which first appeared in [9], are specific examples of multivariable weighted shifts. An excellent survey of the theory of (one variable) weighted shifts, which motivates these maps, can be found in [13]; the corresponding multivariable theory is developed in [7]. It is known that S and T above are commuting subnormals without c.n.e., and also possess the stronger property that $S+T$ and ST fail to be subnormal (in fact, fail to be hyponormal) [10]. The latter provides an answer to a question posed in [5]. We now show that S and T provide an answer to our main question.

For $n = 0,1,\ldots,$ let $H_n = \{f \in H : f(z,w) = z^n g(w)\}$. Then H_n reduces T, S maps H_n into H_{n+1}, and $H = \oplus\Sigma H_n$. Further, $\{z^n w^m : m = 0,1,\ldots\}$ is an orthogonal basis for H_n, and with respect to this basis, $T \mid H_n$ is a (one variable) subnormal weighted shift. By a theorem of C. Berger [4], we have that the m.n.e. of T is given by multiplication by w on $K = \oplus\Sigma K_n$ where

$$K_n = z^n L^2(\mu_n(w)),\ H_n = z^n H^2(\mu_n(w)),$$

and

$$d\mu_n(w) = d\nu_n(r) \times (2\pi)^{-1}\, d\theta \quad \text{where } w = re^{i\theta}$$

$$\int_0^1 r^{2m}\, d\nu_n(r) = \beta^2(n,m) = n!\, m!/(n+m)!, \quad n,m = 0,1,2,\ldots .$$

There is an obvious candidate for S_1, the extension of S to K, namely S_1 defined by $S_1(z^n w^m \bar{w}^p) = z^{n+1} w^m \bar{w}^p$, i.e., multiplication by z. This equation defines a linear map on a dense linear manifold in K. To show S_1 is a bounded operator, it suffices by orthogonality to consider S_1 mapping K_n into K_{n+1}, $n = 0,1,\ldots$.

LEMMA. ν_0 is the unit point mass at $t = 1$ and $d\nu_n(t) = 2nt(1-t^2)^{n-1}\, dt$,

$n = 1,2,\ldots$.

Therefore, $d\nu_{n+1}(t) \leq (n+1)/n \; d\,\nu_n(t)$, $n = 1,2,\ldots$.

Proof. We have $\int_0^1 t^{2m}\, d\,\nu_n(t) = \beta^2(n,m) = n!\, m!/(n+m)!$.

Since $\beta^2(0,m) = 1$ for all m, it follows that $\nu_0 = \delta_1$, the point mass at 1.

Since $\int_0^1 t^{2m}(2t^{2j-1})dt = 1/(m+j)$ and

$$\beta^2(n,m) = \sum_{j=1}^{n} A_j^{(n)}/(m+j) \quad \text{where} \quad A_j^{(n)} = (-1)^{j-1}\, n!/(j-1)!\,(n-j)!\ ,$$

it follows that $d\,\nu_n(t) = p_n(t)\,dt$ where

$$p_n(t) = \sum_{j=1}^{n} 2A_j^{(n)}\, t^{2j-1} = \sum_{j=1}^{n} (-1)^{j-1}\,[n!/j!\,(n-j)!](2j)t^{2j-1}$$

$$= -\ [(1-t^2)^n]' = 2nt(1-t^2)^{n-1},$$

and the lemma follows.

Now, let $f = f(w,\bar{w}) = f(re^{i\theta})$ be a polynomial in $(w,\bar{w})$ and consider $z^n f \in K_n$, $n = 1,2,\ldots$

$$\|S_1(z^n f)\|_K^2 = \|z^{n+1}f\|_K^2 = \|f\|^2_{L^2(\mu_{n+1})}$$

$$= \int_0^1 [M_f(r)]\, d\nu_{n+1}(r)$$

$$\leq (n+1)/n \int_0^1 [M_f(r)]\, d\nu_n(r)$$

$$= (n+1)/n\ \|f\|^2_{L^2(\mu_n)} = (n+1)/n\ \|z^n f\|_K^2,$$

where $M_f(r) = (2\pi)^{-1}\int_0^{2\pi} |f(re^{i\theta})|^2\, d\theta$.

The measure μ_0 is Lebesgue measure on the unit circle and for $f = \sum_{-\infty}^{\infty} a_n e^{in\theta} \in L^2(\mu_0)$,

$$\|S_1 f\|_K^2 = \|zf\|_K^2 = \|f\|^2_{L^2(\mu_1)}$$

$$= \int_0^1 \left(\sum_n |a_n|^2 r^{2n} \right) (2rdr)$$

$$\leq \sum_n |a_n|^2 = \|f\|^2_{L^2(\mu_0)} = \|f\|_K^2 .$$

Thus, S_1 is a bounded linear operator in K, $\|S_1\| \leq 2$ and S_1 is an extension of S commuting with the m.n.e. of T; since S and T do not have c.n.e., S_1 is not subnormal.

It is easy to see that in the basic lemma, one can add the condition $\|S_1\| = \|S\|$. In our example, $\|S\| = 1$ and $\|S_1\| = 2$. Thus, there remains the following

<u>Question</u>. If S and T are commuting subnormals and S extends to an operator S_1 commuting with the m.n.e. of T, $\|S\| = \|S_1\|$, must S_1 be subnormal?

Once again weighted shifts provide a negative answer. For $J = (j,k)$, let $\beta(J) = 1$ if $j = 0$, $\beta(J) = a_k r^j$ otherwise where $0 < r \leq 1$ and $\int_0^1 t^{2k}\, d\rho(t) = a_k^2$, $k = 0,1,\ldots$ for some probability measure ρ on $[0,1]$, ρ not a point mass at $t = 1$.

Let $H = H^2(\beta) = \{\Sigma f_J z^j w^k : \Sigma |f_J|^2 \beta(J)^2 < \infty\}$ and $Sf = zf$ and $Tf = wf$ on H.

Then $H = \oplus \sum_{n=0}^\infty H_n$, where $H_n = \{z^n f(w) \in H\}$.

Letting $\nu_0 = \delta_1$ be the point mass at 1 and $d\nu_n(t) = r^{2n}\, d\rho(t)$, $n = 1,2,\ldots$, we have for each fixed n, $\int_0^1 t^{2m} d\nu_n(t) = \beta^2(n,m)$, $m = 0,1,\ldots$. Thus, $T \mid H_n$ is a subnormal weighted shift for each n, by Berger's theorem, and hence T is subnormal.

Similarly, $H = \oplus \sum_{m=0}^\infty H^{(m)}$ where $H^{(m)} = \{w^m f(z) \in H\}$. Letting $\nu^{(m)} = (1 - a_m^2)\, \delta_0 + a_m^2 \delta_r$, where δ_0 is the point mass at $t = 0$ and δ_r the point mass at $t = r$, we have for each m,

$$\int_0^1 t^{2n} d\nu^{(m)}(t) = \beta^2(n,m), \qquad n = 0,1,\ldots .$$

Thus, S and T are commuting subnormals. We have, as before, the multiplication by z gives an extension S_1 of S that commutes with the m.n.e. of T. Since

$d\nu_{n+1}(t) \leq d\nu_n(t)$ for $n = 1,2,\ldots,$ we have $\|S_1\| = \|S\|$.

To establish our example, it remains to show that S and T do not have c.n.e. By [10], it suffices to show there does not exist a probability measure ν on $[0,1] \times [0,1]$ with $\iint s^{2m}t^{2n}d\nu(s,t) = \beta^2(n,m)$ for $n,m = 0,1,\ldots$. Suppose such a ν exists. Setting $n = 0$, we see ν is supported on the line $s = 1$. Setting $m = 0$, we have, since $a_0 = 1$, that ν must be the unit point mass at $(1,r)$. However, this is a contradiction since $a_m \neq 1$ for $m = 1,2,\ldots$. Thus, S and T do not have c.n.e. and S_1 is therefore not subnormal.

Concerning commutant lifting problems, C. Berger and M. B. Abrahamse raise the question of whether every element in the double commutant of a subnormal operator must lift to the commutant of the m.n.e. We note that the answer to the question is negative. A simple counterexample can be constructed as a sum (non-orthogonal) of two bilateral (one variable) weighted shifts. These correspond to holomorphic functions on overlapping annuli. A somewhat similar example has been given independently by R. Olin and J. Thomson. Details will appear elsewhere. [12].

We conclude with the following somewhat vague

<u>Problem</u>. Find a condition (C) intrinsic to the operators S and T, not involving a positivity condition, such that two commuting subnormals satisfying (C) have c.n.e.

Some possibilities for (C) may be:

(C_1): $p(S,T)$ is subnormal for every polynomial p.

(C_2): $(S + rT)$ is subnormal for all scalars r.

(C_3): $(S + T)$ is subnormal.

By an extension of a result of Lambert [8,11], (C_i) works for general subnormals if and only if it works for two-variable weighted shifts. Recent work of Deddens, however, casts doubt on the plausibility of the C_i.

REFERENCES

[1] M. B. Abrahamse, <u>Commuting subnormal operators</u>, Illinois J. Math., (to appear).

[2] J. Bram, <u>Subnormal operators</u>, Duke Math. J., 22 (1955), 75-94.

[3] P. R. Halmos, Normal dilations and extensions of operators, Summa Brasil. Math., 2 (1950), 125-135.

[4] P. R. Halmos, Ten problems in Hilbert space, Bull. Amer. Math. Soc., 76 (1970), 887-933.

[5] P. R. Halmos, Some unsolved problems of unknown depth about operators on Hilbert space, Proc. Royal Sci. Edin., 76A (1976), 67-76.

[6] T. Ito, On the commutative family of subnormal operators, J. Fac. Sci. Hokkaido Univ., 14 (1958), 1-15.

[7] N. P. Jewell and A. Lubin, Commuting weighted shifts and analytic function theory in several variables, (preprint).

[8] A. Lambert, Subnormality and weighted shifts, J. London Math. Soc., 14 (1976), 476-480.

[9] A. Lubin, Models for commuting contractions, Michigan Math. J., 23 (1976), 161-165.

[10] A. Lubin, Weighted shifts and products of subnormal operators, Indiana U. Math. J., 26 (1977).

[11] A. Lubin, Weighted shifts and commuting normal extension, J. Austr. Math. Soc., (to appear).

[12] A. Lubin, Lifting subnormal double commutants, Studia Math., (to appear).

[13] A. L. Shields, Weighted shift operators and analytic function theory, Math. Surveys 13, Amer. Math. Soc., 1974.

ILLINOIS INSTITUTE OF TECHNOLOGY

NON-SELF-ADJOINT CROSSED PRODUCTS

M. McAsey, P. Muhly and K.-S. Saito

Crossed products arose in algebra during the teens and twenties for the purpose of classifying finite dimensional algebras over arbitrary fields. Murray and von Neumann introduced crossed products into operator theory in their first paper on rings of operators for the purpose of constructing finite factors of type II. Their construction is commonly known as the group-measure construction. From the beginning, in both pure algebra and operator theory, the theory of crossed products has been vigorously pursued not only for the purpose of constructing examples but also for the purpose of developing a general structure theory for broad classes of algebras.

In operator theory, most of the crossed products studied are self-adjoint; i.e., they are either C^*-algebras or von Neumann algebras. In the late sixties, Arveson initiated the study of non-self-adjoint crossed products with at least two objectives in mind. First of all they provide numerous, tractable examples of "non-commutative function algebras" which are amenable to the sort of operator-valued function theory invented by Helson and Lowdenslager, Wiener and Masani, and others, for use in prediction theory and related subjects. It was hoped that these crossed products would point the way to some sort of structure theory for non-self-adjoint operator algebras in general. Secondly, certain of the crossed products Arveson considered turn out to classify ergodic measure preserving transformations up to conjugacy. That is, to each such transformation there is associated a non-self-adjoint crossed product and two transformations are conjugate if and only if the associated crossed products are isomorphic algebras. For these reasons, as well as for others, non-self-adjoint crossed products merit study. We report here on some of our recent investigations into the invariant subspace structure and ideal structure on these algebras.

Throughout this report, M will denote a fixed von Neumann algebra acting on a Hilbert space $\mathfrak{H}$ and u will be a unitary operator on $\mathfrak{H}$ satisfying $uMu^* = M$;

i.e., via conjugation, u implements an automorphism of M. On the Hilbert space $\ell^2(\mathbb{Z}) \otimes \mathfrak{H}$, which we call L^2 and which we frequently identify with $\{f : \mathbb{Z} \to \mathfrak{H} \mid \sum \|f(n)\|^2_{\mathfrak{H}} < \infty\}$, we define the following operators:

$$L_x = I \otimes x, \quad x \in M;$$

$$L_\delta = S \otimes u, \quad S = \text{bilateral shift on } \ell^2(\mathbb{Z});$$

$$R_\delta = S \otimes I; \quad \text{and}$$

R_x, $x \in M'$, is defined by

$$(R_x f)(n) = u^n x u^{*n} f(n).$$

By definition, the von Neumann algebra crossed product determined by M and the automorphism implemented by u is the von Neumann algebra $\mathfrak{L}$ on L^2 generated by $\{L_x\}_{x\in M}$ and L_δ. Similarly, we define $\mathfrak{R}$ to be the von Neumann algebra generated by $\{R_x\}_{x\in M'}$ and R_δ. It is an easy matter to check that $\mathfrak{L}' = \mathfrak{R}$.

We note in passing that the reason for the notation is that we usually assume that M is standard. Under this assumption so are $\mathfrak{L}$ and $\mathfrak{R}$; i.e., they are the left and right algebras of a Hilbert algebra. For the purpose of this report, we don't need the theory of Hilbert algebras, but we find the notation convenient.

By definition, the non-self-adjoint crossed product $\mathfrak{L}_+$ determined by M and the automorphism implemented by u is the weakly closed algebra generated by $\{L_x\}_{x\in M}$ and L_δ. We define $\mathfrak{R}_+$ similarly. Thus, a typical generator for $\mathfrak{L}$ is a sum of the form $\sum_{n=-\infty}^{\infty} L_{x_n} L_\delta^n$ where all but finitely many x_n are zero, while such a sum is in $\mathfrak{L}_+$ if and only if $x_n = 0$ when n is negative. It should be emphasized and kept in mind that the operators L_x and L_δ don't generally commute, rather the equation $L_\delta L_x = L_{uxu^*} L_\delta$ is satisfied. Thus, in a sense, $\mathfrak{L}_+$ may be viewed as an operator-theoretic generalization of a twisted polynomial ring.

A subspace $\mathfrak{M}$ of L^2 is: <u>invariant</u>, if $\mathfrak{L}_+\mathfrak{M} \subseteq \mathfrak{M}$; <u>reducing</u>, if $\mathfrak{L}\mathfrak{M} \subseteq \mathfrak{M}$; <u>pure</u>, if it is invariant and contains no reducing subspaces; and <u>full</u>, if the smallest reducing subspace containing $\mathfrak{M}$ is L^2. We write H^2 for $\{f \in L^2 \mid f(n) = 0, n < 0\}$. Our objective is to prove the following theorem which generalizes the Beurling-Lax-Halmos theorem. In it, we assume that M' is finite and that the automorphism implemented by u preserves a faithful normal finite trace. This implies that $\mathfrak{R}$ is finite. Indeed, if φ_0 is a finite normal trace on M', then φ, defined by $\varphi(\sum R_{x_n} R_\delta^n) = \varphi_0(x_0)$, extends to a finite normal trace on $\mathfrak{R}$ which is faithful if φ_0 is.

THEOREM. If M is a factor and if $\mathfrak{M}$ is a pure invariant subspace, then there is a partial isometry $V \in \mathfrak{R}$ such that $\mathfrak{M} = VH^2$. If $\mathfrak{M}$ is contained in H^2 and invariant under $\mathfrak{R}_+$ as well, then $\mathfrak{M}$ is full and V is unitary.

There is a converse assertion whose proof is too complicated to include here. It states that if M' is finite and if every subspace of H^2 invariant under $\mathfrak{L}_+$ and $\mathfrak{R}_+$ has the indicated form, then M is a factor.

<u>Proof</u>. Let $\mathfrak{M}$ be a pure invariant subspace, let p be the projection onto $\mathfrak{M} \ominus L_\delta \mathfrak{M}$, let q be the projection onto $H^2 \ominus L_\delta H^2$, and write $\mathfrak{L}(M)$ (resp. $\mathfrak{R}(M)$) for $\{L_x\}_{x \in M}$ (resp. $\{R_x\}_{x \in M'}$). Since $\mathfrak{L}(M)$ is self-adjoint, $\mathfrak{M}$ reduces $\mathfrak{L}(M)$; also, since L_δ normalizes $\mathfrak{L}(M)$, $L_\delta \mathfrak{M}$ reduces $\mathfrak{L}(M)$. It follows that $p \in \mathfrak{L}(M)'$, and likewise so is q. Since M is a factor, so is $\mathfrak{L}(M)'$. Consequently p and q are comparable. This means that either there is a partial isometry $v \in \mathfrak{L}(M)'$ such that $vv^* = p$ and $v^*v \leq q$ or there is a partial isometry $v \in \mathfrak{L}(M)'$ such that $vv^* = q$ and $v^*v \leq p$. In the first case, observe that since $\mathfrak{M}$ and H^2 are pure, the projection onto $\mathfrak{M}$ is $\sum_{n=0}^{\infty} L_\delta^n p L_\delta^{*n}$ while the projection onto H^2 is $\sum_{n=0}^{\infty} L_\delta^n q L_\delta^{*n}$. We then define $V = \sum_{n=-\infty}^{\infty} L_\delta^n v L_\delta^{*n}$ to obtain a partial isometry in $\mathfrak{L}' = \mathfrak{R}$ such that $\mathfrak{M} = VH^2$. If the other alternative occurs, then, reversing the roles of p and q, we find a partial isometry V in $\mathfrak{R}$ such that $H^2 = V\mathfrak{M}$. Since H^2 is full and V commutes with L_δ, it follows that V is onto; i.e., V is a co-isometry. But $\mathfrak{R}$ is finite, because M' is, and so V is unitary. Thus $\mathfrak{M} = V^*H^2$. This proves the first assertion.

To prove the second, suppose that $\mathfrak{M} = VH^2$ is contained in H^2 for some non-zero partial isometry V in $\mathfrak{R}$, suppose that $\mathfrak{R}_+\mathfrak{M} \subseteq \mathfrak{M}$, and let e be the final projection of V. Since $\mathfrak{R}$ is finite, we need only prove that $e = I$. Now $eL^2 = \bigvee_{n \leq 0} L_\delta^n \mathfrak{M}$ is invariant under $\mathfrak{R}_+$. Hence e commutes with $\mathfrak{R}(M)$ and $R_\delta e R_\delta^* \leq e$. But $\mathfrak{R}$ is finite, so e commutes with R_δ as well. Thus $e \in \mathfrak{R}'$; and since $e \in \mathfrak{R}$ to begin with, $e \in \mathfrak{B}(\mathfrak{R})$. Once more, since $\mathfrak{R}$ is finite, the initial projection of V is e too. Now as V was constructed, V is what

Halmos calls "rigid analytic". This implies that $eH^2 \subseteq H^2$. Consequently, $e \in \mathfrak{R}_+ \cap \mathfrak{R}_+^* = \mathfrak{R}(M)$. Since $\mathfrak{R}(M)$ is a factor and $e \neq 0$, we conclude that $e = I$ as was to be shown.

When M is not a factor, the invariant subspace structure of $\mathfrak{L}_+$ is highly ramified. The only case for which results have been obtained so far is that which occurs when M is a m.a.s.a. and even here the results are not definitive. However, McAsey has proved the following theorem. Suppose M is a m.a.s.a. and identify it with $L^\infty(\Omega)$ for some probability space Ω. The automorphism of M induced by u is also implemented by a measure preserving transformation τ, say, on Ω. Assume that τ is ergodic. The von Neumann algebra $\mathfrak{L}$, then, is a factor and an example of the Murray-von Neumann group-measure construction. The algebra $\mathfrak{L}_+$ is closely related to the algebras studied by Arveson and is a complete set of conjugacy invariants for τ.

THEOREM. The subspaces $\mathfrak{m}$ of L^2 invariant under both $\mathfrak{L}_+$ and $\mathfrak{R}_+$ are in one-to-one correspondence with subsets E of $\mathbb{Z} \times \Omega$ which are invariant under λ and ρ where λ is defined by the formula $\lambda(n,\omega) = (n+1,\tau\omega)$ and ρ by the formula $\rho(n,\omega) = (n+1,\omega)$. In fact, if L^2 is identified with $L^2(\mathbb{Z} \times \Omega)$ in the obvious way, then $\mathfrak{m} = \{f \in L^2 \mid f$ is supported on $E\}$.

UNIVERSITY OF IOWA

SOME OPERATORS ON $L^2(dm)$ ASSOCIATED WITH FINITE BLASCHKE PRODUCTS

John N. McDonald

We will use A to denote the disk algebra, i.e., the algebra of functions which are continuous on the closed unit disk and analytic on its interior. It is assumed that A is equipped with the sup-norm. We will use L^p to denote $L^p(\Gamma,m)$, where m is normalized Lebesgue measure on the unit circle Γ, and we will use H^p to denote the usual Hardy space; i.e., $H^p = \{f \in L^p \mid \int Z^n f dm = 0\}$, where Z is the identity on Γ. (See [1] for an account of the theory of H^p spaces.) A function $k \in H^\infty$ is called inner if $|k| = 1$ a.e. Inner functions of the form $k = e^{ia} \prod_{j=1}^N (Z - \alpha_j)(1 - \bar{\alpha}_j Z)^{-1}$ where $|\alpha_j| < 1$ for $1 \leq j \leq N$ are called finite Blaschke products. (The finite Blaschke products are exactly the inner functions in A.) All inner functions considered here are non-constant.

We denote by P_A the set of linear operators from A to itself which have norm one and fix the constant functions. The problem of finding the extreme points of P_A was posed by Phelps in [7]. While it might be expected that the extreme points of P_A are multiplicative, in fact, Lindenstrauss, Phelps, and Ryff have given in [2] a specific example of a non-multiplicative extreme element of P_A. (It is easy to show that a linear operator from A to itself is multiplicative if and only if it is a composition operator.) The example found in [2] belongs to a class C of operators on A which satisfy a certain local multiplicative condition, namely,

$$C = \{T \in P_A \mid T(Ff) = GTf \text{ for every } f \in A, \text{ where } F \text{ and } G \text{ are finite Blaschke products}\}.$$

The class C has been studied by Rochberg in [8] and by McDonald in [4] and [5]. In this lecture we discuss an analogous class C' of operators on L^2. In particular, we show that certain results from [8] and [4] are, with modifications, valid for the class C'. We are interested in the class C' because we think it may have significance in the study of composition operators on L^2. Indeed, C' is a natural extension of the class of composition operators. Furthermore, if F is a finite Blaschke product, the proof of Theorem 1 and the remark which follows it yield a family of bounded left inverses for the composition operator $C_F f = f \circ F$. (See Examples 1 and 3.)

For $f,g \in H^\infty$, we denote by $M(f,g)$ the set of bounded linear operators T on L^2 which satisfy the following conditions:

(1) $Tf = g$

(2) $T(fh) = TfTh \ \forall\, h \in L^2$

(3) $T(\overline{h}) = \overline{Th} \ \forall\, h \in L^2$

(4) $T(H^2) \subset H^2$.

The class $\mathbf{C}'$ is the union of the $M(f,g)$'s. An operator T satisfying (1)-(4) might perhaps be described as an analytic local composition operator on L^2.

Example 1.

Suppose that K is non-constant, lies in H^∞, and has absolute value $|K| \leq 1$ a.e. on Γ. Then the composition operator C_K induced by K is bounded [9]. It follows from the fact that the polynomials in Z and $\overline{Z}$ are dense in L^2 and from the continuity of C_K, that $M(Z,K) = \{C_K\}$.

Example 2.

Let λ be a primitive n-th root of unity, where $n \geq 2$. Define an operator T on L^2 by

$$Tf(\exp(i\theta)) = n^{-1} \sum_{k=0}^{n-1} f\left(\lambda^k \exp(i\theta n^{-1})\right).$$

It is straightforward to show that $T \in M(Z^n,Z)$. Clearly, T is not a composition operator.

Example 3.

Let $J,K \in H^\infty$ be non-constant with $|J|,|K| \leq 1$ a.e. on Γ. Suppose that $T \in M(J,K)$, then

(5) $T(u(q \circ J)) = (Tu)(q \circ K)$

for each $u \in L^\infty$ and each $q \in L^2$. In particular, each $T \in M(J,Z)$ is a left inverse for C_J. (We use here the fact that, if $T \in M(J,Z)$, then, by (1) and (2), $T1 = 1$.)

We now consider two finite Blaschke products $F = \Pi_{j=1}^{n}(Z-\alpha_j)(1-\bar{\alpha}_j Z)^{-1}$ and $G = \Pi_{k=1}^{m}(Z-\beta_k)(1-\bar{\beta}_k Z)^{-1}$, where $n,m \geq 1$ and the α_j's and β_k's are (not necessarily distinct) constants of modulus less than 1. Our main result concerns the dimension of the set $M(F,G)$.

THEOREM 1. $M(F,G)$ is a real hyper-plane having dimension $(n-1)(m+1)$.

Before giving a proof of Theorem 1, we will establish some notation and discuss a result due to R. Rochberg. Let $F_0 = 1$, and $F_j = \Pi_{i=1}^{j}(Z-\alpha_i)(1-\bar{\alpha}_i Z)^{-1}$ for $j = 1,2,\ldots,n$. Let $f_j = F\bar{F}_j$ for $j = 0,1,\ldots,n$. Let functions G_k and g_k be defined similarly for $k = 0,1,\ldots,m$. The F_j's form a linearly independent set of functions as do the G_k's. Furthermore, it can be shown that there are non-singular constant matrices $Q_F = (a_{ij})$ and $Q_G = (b_{hk})$ such that

$$f_i = \sum_{j=0}^{n} a_{ij} F_j$$

$$g_h = \sum_{k=0}^{m} b_{hk} G_k$$

for $i = 0,1,\ldots,n$; $h = 0,1,\ldots,m$.

THEOREM 2. (Rochberg [8, Th. 1].) For $i = 0,1,\ldots,n-1$, there exist bounded linear operators P_i on H^2 such that

$$(6) \qquad f = \sum_{j=0}^{n-1} F_j(P_j f) \circ F \;\forall f \in H^2 .$$

Furthermore, if $f \in H^2$ is of the form $f = \Sigma_{j=0}^{n-1} F_j(p \circ F)$, where $p_0, p_1, \ldots, p_{n-1} \in H^2$, then $p_i = P_i f$ for $i = 0,1,\ldots,n-1$.

Rochberg's result is nicely illustrated by the case $F = Z^2$, where

$$P_0 f(\exp(i\theta)) = 2^{-1}(f(\exp(i\theta 2^{-1})) + f(-\exp(i\theta 2^{-1})))$$

$$P_1 f(\exp(i\theta)) = (2\exp(i\theta 2^{-1}))^{-1}(f(\exp(i\theta 2^{-1})) - f(-\exp(i\theta 2^{-1}))) \quad \text{for } \theta \in [0,2\pi).$$

It follows from (5) and (6) that, if $f \in H^2$ and $T \in M(F,G)$ then

$$(7) \qquad Tf = \sum_{j=0}^{n-1} (TF_j)(P_j f) \circ G .$$

Proof of Theorem 1: Let $T \in M(F,G)$. We will show first that there is a constant matrix $M_T = (t_{jk})$ such that

$$(8) \qquad TF_j = \sum_{k=0}^{m} t_{jk} G_k .$$

By (4), $TF_j \in H^2$. Hence, by Theorem 2, there exist constants t_{jk} and functions $q_j \in H^2$ such that, for $j = 0,1,\ldots,n$,

$$TF_j - \sum_{k=0}^{m-1} t_{jk} G_k = q_j G .$$

It follows that

$$G\overline{TF_j} - \sum_{k=0}^{m-1} \overline{t}_{jk} g_k = \overline{q}_j .$$

By (3) and (2), we have

$$Tf_j - \sum_{k=0}^{m-1} \overline{t}_{jk} g_j = \overline{q}_j .$$

Hence, $\overline{q}_j \in H^2$ for $j = 0,1,2,\ldots,n$. Since $H^2 \cap \overline{H}^2$ consists of the constant functions, it follows that the q_j's are constants, which we label t_{jm}. Thus, (8) is valid. Note that, because of the linear independence of the G_k's, the matrix M_T is unique. ([4, Lemma 3.4] gives an explicit method for calculating the t_{jk}'s.) On the other hand, the matrix M_T completely determines T. For, if $f \in H^2$, then (7) and (8) imply,

$$Tf = \sum_{j=0}^{n-1} P_j f \circ G\left(\sum_{k=0}^{m} t_{jk} G_k\right),$$

while, if $f \in (H^2)^{\perp} = \{\overline{g} \mid g \in H^2 \text{ and } \int g dm = 0\}$, then by (3)

$$\overline{Tf} = \sum_{j=0}^{n-1} P_j\bar{f}\circ G\left(\sum_{k=0}^{m} t_{jk}\, G_k\right).$$

Note that the 0th and n^{th} rows of M_T are $(10\cdots 0)$ and $(0\cdots 01)$ respectively. The crucial property of M_T is the following

(9) $$Q_F M_T = \bar{M}_T Q_G .$$

For $i = 0,1,\ldots,n$, we have

$$Tf_i = \sum_{j=0}^{n} a_{ij}\, TF_j = \sum_{k=0}^{n} G_k \sum_{j=0}^{n} a_{ij}\, t_{jk} .$$

On the other hand, by (3), (1), and (2) we have

$$Tf_i = T\bar{F}_i = G \sum_{h=0}^{m} \bar{t}_{ih}\, \bar{G}_h = \sum_{h=0}^{m} \bar{t}_{ih}\, g_h$$

$$= \sum_{k=0}^{m} G_k \sum_{h=0}^{m} \bar{t}_{ih}\, b_{hk} .$$

Equation (9) now follows from the linear independence of the G_k's. Let $M^1(F,G)$ denote the collection of $(n+1)\times(m+1)$ complex matrices which have 0th row $(10\cdots 0)$, n^{th} row $(0\cdots 01)$, and satisfy equation (9). It is not hard to show that $M^1(F,G)$ is a real hyper-plane in the space of all $(m+1)\times(m+1)$ complex matrices, i.e., $M^1(F,G)$ is the translate of a set which is closed under addition and multiplication by real scalars. Furthermore, a technical argument shows that the real dimension of $M^1(F,G)$ is $(n-1)(m+1)$. (See [4, Prop. 3.1].) We have established the existence of a one-to-one mapping $\Psi : M(F,G) \to M^1(F,G)$ given by $\Psi(T) = M_T$. It is not hard to show that Ψ preserves convex combinations. Thus, to finish the proof of Theorem 1, it is enough to show that Ψ is onto. Let $M = (s_{jk}) \in M^1(F,G)$. Define an operator S on L^2 by letting

(10) $$Sf = \sum_{j=0}^{n-1} (P_j f)\circ G \sum_{k=0}^{m} s_{jk}\, G_k$$

for $f \in H^2$ and then extending to all of L^2 via the relation $\overline{Sf} = S\bar{f}$. It is clear that S satisfies (3), (4) and (1) and that

$$SF_j = \sum_{k=0}^{m} t_{jk}\, G_k .$$

Hence, if S can be shown to satisfy (2), it will follow that $\Psi(S) = M$. To verify

$$S(Fu) = G\, Su \ \forall\, u \in L^2,$$

it suffices to consider separately the cases $u \in H^2$ and $u \in \overline{H}^2$. By Theorem 2, the fact that the polynomials in Z are dense in H^2, and the continuity of the operators C_F and C_G, it is enough to consider the cases $u = F_j F^k$ and $u = \overline{F}_j \overline{F}^k$ for $j = 0,1,\ldots,n-1$ and $k = 0,1,\ldots$. If $u = F_j F^k$ then by (10)

$$S(Fu) = S(F_j F^{k+1}) = G^{k+1} \sum_{k=0}^{m} s_{jk}\, G_k = G\, Su .$$

If $u = \overline{F}_j \overline{F}^k$ and $k \geq 1$, then

$$S(Fu) = S(\overline{F}_j \overline{F}^{k-1}) = \overline{S(F_j F^{k-1})} = \overline{G}^{k-1} \sum_{k=0}^{m} \overline{s}_{jk}\, \overline{G}_k .$$

On the other hand,

$$G\, Su = G\, \overline{S(F_j F^k)} = G\, \overline{G}^k \sum_{k=0}^{m} \overline{s}_{jk}\, \overline{G}_k .$$

Thus, $G\, Su = S(Fu)$. If $u = \overline{F}_j$, then

$$S(Fu) = Sf_j = \sum_{i=0}^{n} a_{ji}\, SF_i = \sum_{k=0}^{m} G_k \sum_{i=0}^{n} a_{ji}\, t_{ik} .$$

On the other hand,

$$G\, Su = G\, \overline{SF_j} = G \sum_{h=0}^{m} \overline{t}_{jk}\, \overline{G}_h = \sum_{h=0}^{m} \overline{t}_{jh}\, g_h = \sum_{k=0}^{m} G_k \sum_{h=0}^{m} \overline{t}_{jh}\, b_{hk} .$$

It follows from (9) that $G\, Su = S(Fu)$. Thus, the proof is complete.

REMARK. Explicit expressions can be obtained for the operators P_i, $i = 0,1,\ldots,n-1$. Suppose that f lies in H^2 and is continuous on Γ, such f's are dense in H^2. For $w \in \Gamma$, there are n distinct points $z_0, z_1, \ldots, z_{n-1}$ such that $F(z_i) = w$. Furthermore the matrix $C(w)$ having $i-j$ entry $F_j(z_i)$ is invertible. (See [4].)

Let

$$V_f = \begin{pmatrix} f(z_0) \\ \cdot \\ \cdot \\ \cdot \\ f(z_{n-1}) \end{pmatrix}$$

and

$$U_f = \begin{pmatrix} P_0 f(w) \\ \cdot \\ \cdot \\ \cdot \\ P_{n-1} f(w) \end{pmatrix}$$

By Theorem 2, we have

$$U_f = (C(w))^{-1}\, V_f . \tag{11}$$

Equation (11) can be rewritten

$$U_f = C(w)^*(C(w)C(w)^*)^{-1}\, V_f , \tag{12}$$

where $C(w)^*$ denotes the adjoint of $C(w)$. Let the $i-j$ entry of $(C(w)C(w)^*)^{-1}$ be denoted by c_{ij}. In [5], the c_{ij} are calculated and shown to be rational functions of w and $\bar{w}$. From (12) we have

$$P_i f(w) = \sum_{j=0}^{n-1} \sum_{h=0}^{n-1} c_{jh}\, F_i(z_j) f(z_h) .$$

<u>Questions</u>. Our first question is a general one. Do the operators of the type considered here have any significance for the study of composition operators? We have noted already that the operators in $M(F,Z)$ are left inverses for C_F. Are

there other relationships between the members of $M(F,Z)$ and C_F? Let K be inner. Nordgren has shown that C_K is an isometry iff $\int K dm = 0$. (See [6].) We conjecture that if $T \in M(F,G)$ is an isometry, then T is a composition operator. The conjecture holds if $G = Z$. For if $G = Z$ then T is onto. Thus, if T is an isometry, it must be unitary, and it follows that $C_F = T^{-1}$. It is easy to show that C_F is invertible iff $F = bZ$ where $|b| = 1$. It is now straightforward to show that $Tf(a) = f(\bar{b}z)$. It follows from [3, Th. 1.2], that, if $T \in M(F,G)$, and, if

$$\|Tf\|_\infty = \|f\|_\infty \qquad (\|\cdots\|_\infty = \text{sup-norm})$$

for every bounded continuous f lying in H^2, then T is a composition operator.

REFERENCES

[1] P. Duren, "Theory of H^p Spaces," Academic Press, New York, 1970.

[2] J. Lindenstrauss, R. Phelps and J. V. Ryff, Extreme non-multiplicative operators Lecture notes, University of Washington, Seattle.

[3] J. N. McDonald, Isometries of the disk algebra, Pacific J. Math., 58 (1975), 143-154.

[4] ___________, Convex sets of operators on the disk algebra, Duke Math. J., 42, No. 4 (1975), 787-796.

[5] ___________, Positive operators on the disk algebra, Indiana Univ. Math. J. (to appear).

[6] E. A. Nordgren, Composition operators, Canad. J. Math., 20 (1968), 442-449.

[7] R. R. Phelps, Extreme positive operators and homomorphisms, Trans. Amer. Math. Soc., 108 (1963), 265-274.

[8] R. Rochberg, Linear maps of the disk algebra, Pacific J. Math., 44 (1973), 337-354.

[9] J. V. Ryff, Subordinate H^p functions, Duke Math. J., 33 (1966), 347-354.

ARIZONA STATE UNIVERSITY

A CONCRETE REPRESENTATION OF INDEX THEORY IN VON NEUMANN ALGEBRAS

Catherine L. Olsen*

The object of this paper is to define a natural analytic index function for an arbitrary von Neumann algebra, relative to an arbitrary closed two-sided ideal. This index enables us to develop a complete Fredholm and semi-Fredholm theory in this setting. A concrete representation of the index group for the algebra and the given ideal is obtained as a group of continuous functions on the maximal ideal space of the center, or as a group of germs of such functions. A maximal domain of continuity for this index is discussed. Constructions used in defining the theory are described here, and outlines of proofs of theorems are given; a more detailed version will appear elsewhere.

Recall that for a separable Hilbert space $\mathcal{H}$, an operator T belonging to the algebra $\mathcal{B}(\mathcal{H})$ of all bounded linear operators on $\mathcal{H}$ is Fredholm if it has closed range and if the subspaces kernel T and kernel T^* are both finite-dimensional. If $\mathcal{C}$ denotes the ideal of compact operators in $\mathcal{B}(\mathcal{H})$ and $\pi: \mathcal{B}(\mathcal{H}) \to \mathcal{B}(\mathcal{H})/\mathcal{C}$ is the quotient map onto the C^*-quotient (the Calkin algebra) then F. V. Atkinson's theorem asserts that T is Fredholm if and only if $\pi(T)$ is invertible in $\mathcal{B}(\mathcal{H})/\mathcal{C}$ [1]. For a Fredholm operator T, the classical index is the integer $\text{index}(T) = \dim \ker T - \dim \ker T^*$. The index is a homomorphism of the multiplicative semigroup of Fredholm operators onto the additive group $\mathbf{Z}$ of integers. Moreover, the index is invariant under compact perturbations, and two Fredholm operators have the same index if and only if they belong to the same connected component of the open set of Fredholm operators [1]. It follows that the index induces an isomorphism of the group of connected components of the invertible group $\mathcal{B}(\mathcal{H})/\mathcal{C}$ onto $\mathbf{Z}$.

M. Breuer has developed an abstract Fredholm theory for the relatively compact ideal in a properly infinite semifinite von Neumann algebra on a separable Hilbert space [3,4]. Our index is equivalent to Breuer's in this case, and may be regarded

*Research supported in part by National Science Foundation Grant MPS 73-00562-A03

as a concrete representation of his index. The case considered by Breuer is central, and our development of the properties of the index i for compact ideals uses many of the same ideas. Our theory also uses and extends ideas from discussions of indices in [8, 13, 15]. Some applications of von Neumann index theory are given in [5, 6, 9, 12, 14, 17]. Other abstract generalizations of classical index theory are discussed in [2, 7, 18].

In the following description of our theory, let $\mathfrak{A}$ be a von Neumann algebra, acting on a perhaps nonseparable space, and let $\mathfrak{J}$ denote a closed two-sided ideal in $\mathfrak{A}$. An element A in $\mathfrak{A}$ is called <u>Fredholm</u> relative to $\mathfrak{J}$ if $\pi(A)$ is invertible in $\mathfrak{A}/\mathfrak{J}$, where $\pi\colon \mathfrak{A} \to \mathfrak{A}/\mathfrak{J}$ is the quotient map. Although the range of such a Fredholm element need not be closed, there is a natural analog for the classical result of Atkinson described above: A is Fredholm for $\mathfrak{J}$ if the orthogonal projections onto the kernel of A and the kernel of A^* both belong to $\mathfrak{J}$, and if there is a projection E in $\mathfrak{J}$ such that the range of A contains the range of $I - E$ [4]. An element A is called <u>left Fredholm</u> if $\pi(A)$ is left invertible; or <u>right Fredholm</u> if $\pi(A)$ is right invertible; or <u>semi-Fredholm</u> if for some central projection P, PA is left Fredholm in $P\mathfrak{A}$ and $(I - P)A$ is right Fredholm in $(I - P)\mathfrak{A}$.

To define our analog to the classical index, we need a notion of the dimension of a projection in an arbitrary von Neumann algebra, and such a notion has been provided by J. Tomiyama [19]. The center Z of $\mathfrak{A}$ is an abelian von Neumann algebra, its maximal ideal space Ω is hyperstonean, and $Z \simeq C(\Omega)$. To each projection E in $\mathfrak{A}$, Tomiyama associates a continuous function $\dim E$ on Ω. If $\mathfrak{A}$ is of type I or III, the function $\dim E$ will be cardinal-valued, while if $\mathfrak{A}$ is of type II, the values of $\dim E$ will be nonnegative reals or infinite cardinal numbers. The von Neumann algebra $\mathfrak{A}$ has a central decomposition into algebras of types I, II and III, with a corresponding partition of the maximal ideal space $\Omega = \Omega_I \cup \Omega_{II} \cup \Omega_{III}$. Let D denote the set of all infinite cardinal numbers less than or equal to the dimension of $\mathcal{H}$, where $\mathfrak{A} \subseteq \mathfrak{B}(\mathcal{H})$. Following the notation of W. Wils [20], define value sets $V_I = -D \cup \mathbb{Z} \cup D$, $V_{II} = -D \cup \mathbb{R} \cup D$, and $V_{III} = -D \cup \{0\} \cup D$. Give each V_j the order topology, and let V be the disjoint union of the V_j's. Define the set $C_c(\Omega)$ of continuous functions by

$$C_c(\Omega) = \{f \mid f: \Omega \to V \text{ is continuous and } f(\Omega_j) \subset V_j, \text{ each } j\}.$$

Two projections E and F in a von Neumann algebra $\mathfrak{a}$ are said to be equivalent ($E \sim F$), if $E = U^*U$ and $F = UU^*$ for some partial isometry U in $\mathfrak{a}$. Tomiyama showed the existence of a dimension function dim mapping the projections of $\mathfrak{a}$ into the positive functions in $C_c(\Omega)$ such that for any projections E and F:

(i) $0 \leq \dim E \leq$ dimension $\mathcal{H}$, and $\dim E = 0 \Leftrightarrow E = 0$;

(ii) $\dim E \leq \dim F \Leftrightarrow E \sim Q$ for some projection $Q \leq F$;

(iii) E orthogonal to $F \Rightarrow \dim (E + F) = \dim E + \dim F$;

(iv) $\dim PE = P \dim E$, for any central projection P;

(we identify P with the characteristic function of some open and closed subset of Ω). This finitely additive dimension function generalizes the well-known countably additive dimension function for finite type algebras.

By analogy with the classical case, we would like to define the index of an element A in $\mathfrak{a}$ to be a function in $C_c(\Omega)$ given by $\dim N_A - \dim N_{A^*}$, where N_A and N_{A^*} are the projections onto the kernel of A and the kernel of A^*, respectively. However, since these functions are in general infinite-valued, difficulties in cardinal arithmetic arise: the pointwise difference will usually not be a continuous function on Ω. We overcome this problem as follows: given two positive functions f,g in $C_c(\Omega)$, define $f - g$ pointwise on the open set X in Ω where $f \neq g$. Then the closure $\overline{X}$ is open and closed, and is equal to the Stone-Čech compactification of X, so extend $f - g$ continuously to $\overline{X}$. Finally, set $f - g \equiv 0$ on $\Omega \setminus \overline{X}$. Using this subtraction, define the index map $i: \mathfrak{a} \to C_c(\Omega)$ by

$$i(A) = \dim N_A - \dim N_{A^*},$$

for each A in $\mathfrak{a}$.

Keeping in mind that we want the range of the index to have a group structure, we similarly define an operation of addition on all of $C_c(\Omega)$ (define $f + g$ pointwise on the set where $f \neq -g$, extend continuously to the closure of this set, and let $f + g \equiv 0$ off this closure).

It is easy to see that this operation is commutative, has an identity and inverses, but is not in general associative. However, the operation is associative when restricted to the set

$$\{f \in C_c(\Omega): f \text{ is finite on some dense open subset of } \Omega\},$$

which we will call the group of almost everywhere finite functions (finite a.e.).

A projection E in $\mathfrak{a}$ is relatively finite if $E \sim F$ and $F \leq E$ implies $E = F$. It is not hard to show that the relatively finite projections are precisely those E with $\dim E$ finite on a dense open set in Ω. Define the relatively compact ideal $\mathcal{K}$ in $\mathfrak{a}$ to be the ideal generated by the relatively finite projections. We see next that for any ideal contained in $\mathcal{K}$, the map i has the desired properties for an index relative to $\mathcal{J}$.

THEOREM 1. Let $\mathcal{J}$ be any ideal contained in $\mathcal{K}$. On the open set of semi-Fredholm elements for $\mathcal{J}$ the map i is invariant under perturbation by elements of $\mathcal{J}$, i is constant on connected components, and i distinguishes between components.

PROOF. Observe first the extent to which the map i is a homomorphism: that is, $i(AB) = i(A) + i(B)$ whenever both N_A and N_B are relatively finite; or whenever both N_A and $N_A{}^*$ are relatively finite and $i(A) = 0$; or whenever N_{AB} is relatively finite and $i(AB) = 0$. This is established by using properties of equivalence of projections to rewrite the desired conclusion as an equation involving six independent functions in $C_c(\Omega)$; then one checks the cardinal arithmetic to see that this equation holds under the various hypotheses. It is necessary to use the fact that $\dim E$ is finite a.e. on Ω whenever E is a relatively finite projection.

It follows readily from this that for any A in $\mathfrak{a}$ and for any C in $\mathfrak{a}$ with relatively finite range, that $i(A + C) = i(A)$: for,

$$i(A + C) + i(N_C) = i((A + C)N_C) = i(AN_C) = i(A) + i(N_C) .$$

It also follows that i is locally constant on the open set of semi-Fredholm elements: for sufficiently small ϵ, $\|T - S\| < \epsilon$ implies S is semi-Fredholm and $i(T) = i(S)$, whenever T is semi-Fredholm. We sketch this argument in case T is left-Fredholm: choose $\epsilon < m(T)$ where $m(T)$ is the minimum of the spectrum of $\pi((T^*T)^{\frac{1}{2}})$. Define a Δ with relatively finite range by the equation

$$T + \Delta = T(I - E) + \epsilon VE$$

where $T = V|T|$ is the polar decomposition, V is an isometry in $\mathcal{G}$ and $E = E[0,\epsilon]$ is the spectral projection for $|T|$, $E \in \mathcal{J}$. There is a $B \in \mathcal{G}$ with $B(T + D) = I$ and $\|B\| \leq 1/\epsilon$. Thus $B(S + D)$ is also invertible since $\|I - B(S + D)\| < 1$, so S is left Fredholm. Using the first observation of the proof yields

$$i(B) + i(S + D) = i(B(S + D)) = i(B(T + D)) = i(B) + i(T + D)$$

so $i(S) = i(T)$, by the preceding paragraph.

Invariance of i under perturbation by elements of $\mathcal{J}$ now follows immediately: if A is left-Fredholm and $K \in \mathcal{J}$, then approximate K by some C in $\mathcal{J}$ having relatively finite range; so that $\|(A + K) - (A + C)\| < \epsilon$ measures a small difference of left Fredholm elements.

Finally we describe how i distinguishes between components of the set of semi-Fredholm elements. If A and B in $\mathcal{G}$ satisfy $i(A) = i(B) < 0$, we can construct a left invertible path $A_t \subset \mathcal{G}$ from A to B with $i(A_t) = i(A)$: first connect A to U, where U is an isometry for the polar decomposition $A = U|A|$, observing that necessarily $i(A) = i(U)$. Similarly connect V to B, where $B = V|B|$. Let W be a unitary in $\mathcal{G}$ for the polar decomposition $UV^* = W|UV^*|$. There is an invertible path in $\mathcal{G}$ from W to I, and hence an invertible path $B_t \subset \mathcal{G}$ from UV^* to I, and hence B_tV is a left invertible path from U to V. The general semi-Fredholm case follows by a central decomposition, and checking details.

A characterization of the closed ideals in a von Neumann algebra due to W. Wils [20] has proved very useful in developing this theory. Wils shows that the range dim $\mathcal{A}$ of the dimension function is precisely

$$\dim \mathcal{A} = \{f \in C_c(\Omega) : 0 \le f \le \dim(I_{\mathcal{A}})\} .$$

We may restrict our theory to properly infinite algebras (those containing no finite central projection) since it is easy to see that $i(A) = 0$ if there is a finite central projection P with $PA = A$. For such algebras, Wils shows there is a lattice isomorphism between the lattice of closed two-sided ideals in $\mathcal{A}$, and the lattice of order ideals in the positive cone of $C_c(\Omega)$; this correspondence is given by $\mathcal{J} \leftrightarrow \dim \mathcal{J}$.

For any ideal $\mathcal{J}$ in $\mathcal{A}$, define the <u>index group</u> $I(\mathcal{A},\mathcal{J})$ to be the group of connected components of the invertible group of $\mathcal{A}/\mathcal{J}$ (this is isomorphic to the group of components of the set of Fredholm elements for $\mathcal{J}$). We wish now to use the index i to represent the index group for any relatively compact ideal:

THEOREM 2. Let $\mathcal{K}$ be the relatively compact ideal in $\mathcal{A}$, and let $\mathcal{J}$ be an ideal of $\mathcal{A}$ contained in $\mathcal{K}$. The index i is a continuous homomorphism of the multiplicative semigroup of Fredholm elements for $\mathcal{J}$ onto a subgroup of the discrete group of finite a.e. functions in $C_c(\Omega)$, which induces the following isomorphisms:

(i) the index group $I(\mathcal{A},\mathcal{K})$ is isomorphic to the group of finite a.e. functions;

(ii) the index group $I(\mathcal{A},\mathcal{J})$ is isomorphic to the subgroup of finite a.e. functions consisting of $\{f: |f| \in \dim(\mathcal{J})\}$.

PROOF. It follows as a corollary to Theorem 1 and to the observations in its proof that i is a homomorphism on the group of Fredholm elements which is constant on components and distinguishes between components (those components of the set of semi-Fredholm elements on which $|i(A)| \in \dim \mathcal{J}$ consist entirely of Fredholm elements). Thus i induces a one-to-one homomorphism of $I(\mathcal{A},\mathcal{J})$ onto a subgroup of the finite a.e. group. To see this maps onto the desired subset of $C_c(\Omega)$, it is straightforward to construct for each $f \le 0$ in $C_c(\Omega)$, an isometry U in $\mathcal{A}$ with $i(U) = f$. For $f \ge 0$ one gets a coisometry, and there is a central decomposition into these cases.

For an ideal $\mathcal{J}$ not contained in the relatively compact ideal, the function i does not behave like an index: i is not invariant under ideal perturbations, and i is not constant on the connected components of the set of Fredholm elements for $\mathcal{J}$. To deal with this, observe first that there is a central decomposition of $\mathcal{J}$ into a relatively compact summand, and a completely noncompact summand (no central summand of the latter is relatively compact). The theory thereby splits into two distinct cases: relatively compact ideals as taken care of above, and completely noncompact ideals. For each of the latter ideals we must modify the map i by taking a quotient, to obtain an index with all the desired properties.

If $\mathcal{J}$ is a completely noncompact ideal in $\mathcal{A}$, there is a dense open subset Δ of Ω defined by

$$\Delta = \{\lambda \in \Omega: \text{ some } f \text{ in } \dim \mathcal{J} \text{ is infinite on a neighborhood of } \lambda\}.$$

Specify a subset J of $C_c(\Omega)$ by

$$J = \{f \in C_c(\Omega): |f| \in \dim \mathcal{J} \text{ and } f \text{ is zero on some neighborhood of } \Omega \setminus \Delta\}.$$

Then equivalence mod J is a congruence for the operation of addition on $C_c(\Omega)$, so that a corresponding operation is induced on cosets in $C_c(\Omega)/J$. Define the index map

$$i: \mathcal{A} \to C_c(\Omega)/J \text{ by } \bar{i}(A) = i(A) + J$$

for every A in $\mathcal{A}$.

THEOREM 3. Let $\mathcal{J}$ be a completely noncompact ideal of $\mathcal{A}$. On the open set of semi-Fredholm elements for $\mathcal{J}$ the map $\bar{i}$ is invariant under perturbations by elements of $\mathcal{J}$, $\bar{i}$ is constant on connected components, and $\bar{i}$ distinguishes between components.

PROOF. The crucial homomorphism property of $\bar{i}$, that $\bar{i}(AB) = \bar{i}(A) + \bar{i}(B)$, now holds whenever N_A and $N_B \in \mathcal{J}$; or N_A and $N_{A^*} \in \mathcal{J}$ and $\bar{i}(A) = 0$; or $N_{AB} \in \mathcal{J}$ and $\bar{i}(AB) = 0$. This is shown in a fashion analogous to the compact case, by using the finite a.e. representative in each coset. It then follows as for Theorem 1: if A and C belong to $\mathfrak{A}$ with the range projection of C in $\mathcal{J}$ then $\bar{i}(A + C) = \bar{i}(A)$. Also analogous is the argument that $\bar{i}$ is locally constant on the set of semi-Fredholm elements for $\mathcal{J}$, and that $\bar{i}(A + K) = \bar{i}(A)$ whenever A is semi-Fredholm and K is in $\mathcal{J}$.

Again as for Theorem 1 it is possible to construct a semi-Fredholm path A_t in $\mathfrak{A}$ from A to B with $\bar{i}(A_t) = \bar{i}(A)$, whenever $\bar{i}(A) = \bar{i}(B)$. However, the argument is rather more elaborate. By taking a central decomposition, several cases are considered separately: for case (i), assume some f in $\dim \mathcal{J}$ is infinite on Ω so that $\Delta = \Omega$, and assume $i(B) \leq i(A) \leq 0$. Begin as for Theorem 1; then to connect U to V, first connect U to $U(I - E)$ where E is a properly infinite projection in $\mathcal{J}$ with $\dim E$ infinite and such that $\dim E \geq |\dim N_U{}^* - \dim N_V{}^*|$. Choose this path so that i is constant on it. Now connect $U(I - E)$ to $U(I - E) + W$, where W is a partial isometry in $\mathfrak{A}$ such that $W^*W = E$ and $WW^* = UEU^* - F$, with $\dim F = |\dim N_U{}^* - \dim N_V{}^*|$. This path can be chosen so that $\bar{i}$ is constant, although i will jump. Finally, $i(U(I - E) + W) = i(V)$, so we can connect $U(I - E) + W$ to V by a path with i constant.

For case (ii), again assume some f in $\dim \mathcal{J}$ is infinite on Ω, but now require $i(B) \leq -i(A) \leq 0$. Then $\bar{i}(B) = \bar{i}(A) = 0$. Proceed as in case (i), only, connect U^* to I instead of U to V. Then similarly connect V to I. Taking adjoints gives a path from U to I.

The other cases where $\dim \mathcal{J}$ contains an infinite function are similar. If $\dim \mathcal{J}$ contains no such infinite f, so $\Omega \setminus \Delta \neq \phi$, then $i(A) \equiv i(B)$ on some open and closed neighborhood of $\Omega \setminus \Delta$. Thus we can reduce by a central decomposition, to one of the above cases, and to Theorem 1.

In order to represent the index group $I(\mathfrak{A},\mathcal{J})$ for a completely noncompact ideal $\mathcal{J}$, define G and H to be the following subgroups of the group of finite a.e. functions:

$$G = \{f \in C_c(\Omega): |f| \in \dim(\mathcal{J}) \text{ and } f \text{ is finite a.e.}\}$$

$$H = \{f \in G: f \text{ vanishes on a neighborhood of } \Omega \setminus \Delta\}.$$

THEOREM 4. Let $\mathcal{J}$ be a completely noncompact ideal in $\mathcal{C}$. The index $\bar{i}$ is a continuous homomorphism of the multiplicative semigroup of Fredholm elements for $\mathcal{J}$ onto a discrete group of cosets in $C_c(\Omega)/J$, which induces an isomorphism $I(\mathcal{C},\mathcal{J}) \simeq G/H$. That is, $I(\mathcal{C},\mathcal{J})$ is isomorphic to the group of germs at $\Omega \setminus \Delta$ of the functions in G.

PROOF. This proof is precisely analogous to that for Theorem 2. One must observe that although the image under i of the Fredholm elements for $\mathcal{J}$ is not a group of functions in $C_c(\Omega)$, nevertheless the image under $\bar{i}$ is a group of cosets in $C_c(\Omega)/J$. It is not hard to see that each coset $g + J$ in the image contains a finite a.e. function, say g_0, with $|g_0| \in \dim \mathcal{J}$. The group operation can be performed with these representatives.

We see that the index group $I(\mathcal{C},\mathcal{K})$ is central to the theory: for every ideal in $\mathcal{C}$, $I(\mathcal{C},\mathcal{J})$ is isomorphic to a subgroup of $I(\mathcal{C},\mathcal{K})$ or to a quotient of such a subgroup.

A second part of this investigation includes a description of the maximal domain of continuity for the index maps, and a study of the limit points of the components of the set of semi-Fredholm elements. We briefly mention some of these results. The index for an ideal $\mathcal{J}$ extends continuously to an open maximal domain $\mathcal{D}$ strictly larger than the domain of semi-Fredholm elements for $\mathcal{J}$ if and only if $\mathcal{J}$ does not contain the <u>strong radical</u> of $\mathcal{C}$ (the intersection of the maximal ideals of $\mathcal{C}$). There is still an identifiable group of Fredholm components of $\mathcal{D}$, and the index is a homomorphism on these, and on the semi-group of left Fredholm components. If $\mathcal{J}$ contains the strong radical, the index does not extend continuously beyond the semi-Fredholm elements for $\mathcal{J}$; this is true in the classical case where the ideal of compact operators is the strong radical of $\mathcal{B}(\mathcal{H})$ [8]. It is possible to compute the distance from each element of A to each connected component of the open set of semi-Fredholm elements in terms of natural parameters, and to thereby describe the closure of each component. In the classical case, each operator which is not semi-Fredholm is a limit point of every component [8]. In general, some elements may be in the closure of a unique component. In some cases these will be bounded away from the complement of this component, in other cases they will be a limit point of an infinite union of other components. Other elements may be in the closure of many components or of every component.

REFERENCES

[1] F. V. Atkinson, The normal solubility of linear equations in normal space, Math. Sbornik N. S., 28 (70), (1951), 3-14 (Russian).

[2] B. A. Barnes, The Fredholm elements of a ring, Canad. J. Math., 21 (1969), 84-95.

[3] M. Breuer, Fredholm theories in von Neumann algebras I, Math. Ann., 178 (1968), 243-254.

[4] M. Breuer, Fredholm theories in von Neumann algebras II, Math. Ann, 180 (1969), 313-325.

[5] M. Breuer, Theory of Fredholm operators and vector bundles relative to a von Neumann algebra, Rocky Mountain J. Math., 3 (1973), 383-429.

[6] M. Breuer and R. S. Butcher, Fredholm theories of mixed type with analytic index functions, Math. Ann., 209 (1974), 31-42.

[7] S. R. Caradus, W. E. Pfaffenburger and B. Yood, Calkin Algebras and Algebras of Operators on Banach Spaces, Marcel Dekker, New York: 1974.

[8] L. A. Coburn and A. Lebow, Algebraic theory of Fredholm operators, J. Math. and Mech., 15 (1966), 577-583.

[9] L. A. Coburn, R. G. Douglas, D. G. Schaeffer and I. M. Singer, C^*-algebras of operators on a half-space II: index theory, Inst. Haut Etude Sci. Publ. Math., 40 (1971), 69-79.

[10] H. O. Cordes and J. P. Labrousse, The invariance of the index in the metric space of closed operators, J. Math. and Mech., 12 (1963), 693-720.

[11] R. G. Douglas, Banach Algebra Techniques in Operator Theory, Academic Press, New York: 1972.

[12] K. E. Ekman, Indices on C^*-algebras through representations in the Calkin algebra, Duke J. Math., 41 (1974), 413-432.

[13] M. Gartenberg, Extensions of the index in factors of type II_∞, Proc. Amer. Math. Soc., 43 (1974), 163-168.

[14] C. L. Olsen, Approximation by unitary elements in a von Neumann algebra, in preparation.

[15] M. J. O'Neill, Semi-Fredholm operators in von Neumann algebras, University of Kansas Technical Report No. 21 (New Series).

[16] S. Sakai, C^*-algebras and W^*-algebras, Springer-Verlag, New York: 1971.

[17] D. G. Schaeffer, An application of von Neumann algebras to finite difference equations, Ann. Math., 95 (1972), 116-129.

[18] M. R. F. Smyth, Fredholm theory in Banach algebras, Trinity College Dublin preprint, 1975.

[19] J. Tomiyama, Generalized dimension function for W^*-algebras of infinite type, Tohoku Math. J., 10 (1958), 121-129.

[20] W. Wils, Two-sided ideals in W^*-algebras, J. fur die Reine und Angewandte Math., 242-244 (1970), 55-68.

STATE UNIVERSITY OF NEW YORK AT BUFFALO

A CLASSIFICATION PROBLEM FOR ESSENTIALLY n-NORMAL OPERATORS

Norberto Salinas

1. INTRODUCTION

In this paper we present a survey of some of the main results obtained in [16], [10] and [11], and we offer a new viewpoint in regard to the classification of n-essentially normal operators. Our aim has been to try to generalize the results obtained by Brown, Douglas and Fillmore [3] concerning essentially normal operators, taking the approach suggested by Arveson [1] (see also [15]).

In §2 we first remind the reader of the above mentioned Arveson's approach and we explain how this line of argument leads to more transparent proofs of the results obtained in [16]. In §3 we describe two significant concrete examples of essentially n-normal operators, which were presented in [11], and we discuss some of their properties with relation to the results given in §2.

Finally, we would like to thank Vern Paulsen for his helpful comments which enabled us to introduce several improvements in the present revised version of this survey.

2. THE CLASSIFYING STRUCTURE $EN_n(X)$.

Throughout, $\mathcal{H}$ will denote a fixed, separable, infinite dimensional, complex Hilbert space, $\mathcal{L}(\mathcal{H})$ will denote the algebra of all (bounded, linear) operators on $\mathcal{H}$, and $\mathcal{K}(\mathcal{H})$ will denote the ideal of all compact operators on $\mathcal{H}$. Also, the Calkin algebra $\mathcal{L}(\mathcal{H})/\mathcal{K}(\mathcal{H})$ will be denoted by $\mathfrak{Q}(\mathcal{H})$, and $\pi : \mathcal{L}(\mathcal{H}) \to \mathfrak{Q}(\mathcal{H})$ will denote the canonical quotient map.

In what follows, X will denote a finite dimensional compact Hausdorff space; that is, there exists a positive integer m such that $X \subset \mathbf{C}^m$. In this case, the unital C^*-algebra $C(X)$ of all continuous complex-valued functions on X is generated by the coordinate functions χ_j, $1 \leq j \leq m$. Given an m-tuple of essentially commuting, essentially normal operators T_j in $\mathcal{L}(\mathcal{H})$, whose joint essential spectrum is X, the correspondence $\chi_j \to \pi T_j$, $j = 1,\ldots,m$ determines a unital $*$-monomorphism $\tau : C(X) \to \mathfrak{Q}(\mathcal{H})$. The set of these $*$-monomorphisms can be identified with the set of C^*-extensions of $\mathcal{K}(\mathcal{H})$ by $C(X)$ (see [3]). Given two m-tuples of essentially commuting, essentially normal operators T_j, S_j, $j = 1,\dots,m$ we shall say that they are <u>compalent</u>, if there exists a unitary operator U in $\mathcal{L}(\mathcal{H})$ such that $U^*T_jU - S_j \in \mathcal{K}(\mathcal{H})$ for every $j = 1,2,\dots,m$.

Following [1] and [3], we shall denote by $\mathrm{Ext}(X)$ the set of all compalence classes $[T_j]$ of m-tuples of essentially commuting, essentially normal operators T_j whose joint essential spectrum coincides with X. We define the following abelian operation on $\mathrm{Ext}(X)$: Given $[S_j]$ and $[T_j]$ in $\mathrm{Ext}(X)$ we define $[S_j] + [T_j] = [S_j \oplus T_j]$.

The following theorem was proved in [3] employing C^*-algebra extensions and methods from Algebraic Topology. Afterwards, in [1] an operator-theoretic proof of that result was given using dilation theory.

THEOREM 2.1. $\mathrm{Ext}(X)$ is an abelian group. The set of all m-tuples S_j of commuting normal operators whose joint essential spectrum is X is contained in a unique compalence class, which is the neutral element of $\mathrm{Ext}(X)$.

<u>Definitions</u> 2.2.

We shall say that an operator T in $\mathcal{L}(\mathcal{H})$ is (<u>essentially</u>) <u>n-normal</u>, if T is unitarily equivalent to an $n \times n$ matrix of (essentially) commuting, (essentially normal operators on $\mathcal{H}$.

A suitable invariant for essentially n-normal operators that plays the role of the joint essential spectrum for m-tuples of essentially commuting, essentially normal operators was first discussed in [12], and is defined as follows:

<u>Definition</u> 2.3.

Given T in $\mathcal{L}(\mathcal{H})$, let $R_e^n(T) = \{L \in \mathfrak{M}_n :$ there exists a unital *-homomorphism $\varphi : \pi C^*(T) \to \mathfrak{M}_n$ such that $\varphi(\pi T) = L\}$, where $C^*(T)$ denotes the C^*-algebra generated by T and $1_{\mathcal{H}}$, and $\mathfrak{M}_n$ is the C^*-algebra of all $n \times n$ complex matrices

If $X = R_e^n(T)$ for some T in $\mathcal{L}(\mathcal{H})$, then X enjoys the following properties: it is compact, invariant under unitary conjugation in $\mathfrak{M}_n$, and given $L_j \in X$, and projections P_j in $\mathfrak{M}_n$, such that $P_jL_j = L_jP_j$, $1 \le j \le k$, and $\sum_{j=1}^{k} P_j = 1$, then

$\sum_{j=1}^{k} P_j L_j \in X$, $k = 1,\dots,n$. Furthermore, it is easy to check that if T is essentially n-normal, then $R_e^n(T) \neq \emptyset$, [13].

Definition 2.4.

A nonempty subset X of $\mathfrak{M}_n$ enjoying the above properties will be called hypoconvex, and if Y is any nonempty subset of $\mathfrak{M}_n$, the smallest hypoconvex set containing Y will be denoted by $\hat{Y}$. Also, given a hypoconvex set $X \subset \mathfrak{M}_n$, we define the following equivalence relation on the set of essentially n-normal operators T in $\mathcal{L}(\mathcal{H})$ such that $R_e^n(T) = X$: we say that T is weakly compalent to T', and we write $T \approx T'$, if there exists an essentially unitary operator V in $\mathcal{L}(\mathcal{H})$ such that $V^*TV - T' \in \mathcal{K}(\mathcal{H})$. The set of all weak compalence classes $[T]$ of essentially n-normal operators T in $\mathcal{L}(\mathcal{H})$ such that $R_e^n(T) = X$ will be denoted by $EN_n(X)$.

REMARK 2.5. Given a hypoconvex set $X \subset \mathfrak{M}_n$, there exists a natural map $\beta : Ext(X) \to EN_n(X)$. Indeed, if T_{ij}, $1 \le i, j \le n$, is an n^2-tuple of essentially commuting, essentially normal operators, whose joint essential spectrum is X, and (T_{ij}) is the $n \times n$ operator matrix whose entries are T_{ij} (notice, therefore, that (T_{ij}) is an essentially n-normal operator), then we set $\beta[T_{ij}] = [(T_{ij})]$. In order to show that β is well defined we must prove that $R_e^n((T_{ij})) = X$. To this end, let $\mathfrak{A}$ be the unital C^*-algebra generated by T_{ij}, $1 \le i,j \le n$. Then, there exists a unital $*$-monomorphism $\tau : C(X) \to \pi\mathfrak{A}$, uniquely determined by $\tau(X_{ij}) = \pi T_{ij}$, where X_{ij} are the coordinate functions on X, $1 \le i,j \le n$. It follows that $\pi C^*((T_{ij}))$ is $*$-isomorphic to a C^*-subalgebra of $C(X) \otimes \mathfrak{M}_n$. This easily implies that $X \subset R_e^n((T_{ij}))$. To prove the other inclusion, let $L \in R_e^n(T)$, and let $L = \sum_{j=1}^{k} \oplus L_j$ be the decomposition of L into irreducible matrices. Using the $*$-isomorphism between $\pi C^*((T_{ij}))$ and a C^*-subalgebra of $C(X) \otimes \mathfrak{M}_n$, it follows from [5, Prop. 2], that there exists a matrix L_j' such that $L_j \oplus L_j' \in X$, for $1 \le j \le k$. Thus, by the hypoconvexity of X, we deduce that $L \in X$, as desired. In particular, if $\{L_m\}$ is a dense sequence in X, whose terms are repeated infinitely often, then $S = \sum_{m=1}^{\infty} \oplus L_m$ can be identified with an $n \times n$ matrix (S_{ij}) of commuting normal operators, whose joint essential spectrum is X.

This means that S is an n-normal operator such that $R_e^n(S) = X$.

THEOREM 2.6. [16, Thm. 3.8]. The map $\beta : \mathrm{Ext}(X) \to \mathrm{EN}_n(X)$, defined in the above remark, is surjective.

Proof. Let T be an essentially n-normal operator in $\mathcal{L}(\mathcal{H})$ such that $R_e^n(T) = X$, and let $U : \mathcal{H} \to \mathcal{H} \otimes C^n$ $(= \mathcal{H} \oplus \mathcal{H} \oplus \cdots \oplus \mathcal{H})$ be a unitary transformation such that UTU^* is an $n \times n$ matrix (T_{ij}) of essentially commuting, essentially normal operators. From [13, Theorem 4.4], if S is any n-normal operator in $\mathcal{L}(\mathcal{H})$ such that $R_e^n(S) = X$, then $T \oplus S \approx T$. Thus, we let (S_{ij}) be the $n \times n$ matrix of commuting normal operators defined at the end of the above remark, so that $R_e^n((T_{ij}) \oplus (S_{ij})) = X$. On the other hand, it is also clear that the joint essential spectrum of the n^2-tuple $T_{ij} \oplus S_{ij}$ contains X (since so does the joint essential spectrum of the n^2-tuple S_{ij}). To see that the opposite inclusion is valid we simply notice that if ω is any character of the unital C^*-algebra generated by $\pi(T_{ij} \oplus S_{ij})$, $1 \leq i, j \leq n$, then $(\omega\pi(T_{ij} \oplus S_{ij}))$ is an $n \times n$ complex matrix in $R_e^n((T_{ij} \oplus S_{ij})) = R_e^n((T_{ij})) = X$. Therefore, the compalence class of the n^2-tuple $T_{ij} \oplus S_{ij}$ is in $\mathrm{Ext}(X)$ and the theorem is proved.

THEOREM 2.7. [15, Thm. 4]. $\mathrm{EN}_n(X)$ forms an abelian group, with the operation given by $[T] + [T'] = [T \oplus T']$, whose neutral element is the compalence class containing all n-normal operators S such that $R_e^n(S) = X$.

Proof. We first notice that if T and T' are essentially n-normal operators such that $R_e^n(T) = R_e^n(T') = X$, then $R_e^n(T \oplus T') = X$. This follows from [13, Theorem 4.4], Theorem 2.6, and the fact that $\beta([T_{ij}] + [T'_{ij}]) = \beta[T_{ij} \oplus T'_{ij}] = [(T_{ij} \oplus T'_{ij})] = [(T_{ij}) \oplus (T'_{ij})]$. The last observation also implies that β is a homomorphism. Since $\mathrm{Ext}(X)$ is a group (Theorem 2.1) it follows that $\mathrm{EN}_n(X)$ is also a group. Since the last assertion of the theorem follows from [13, Theorem 4.4] the proof is complete.

DEFINITION 2.8. Let X be a hypoconvex set of $\mathfrak{M}_n$ and let $C^*_\chi(X)$ be the unital C^*-subalgebra of $C(X) \otimes \mathfrak{M}_n$ generated by χ, where $\chi : X \to X$ is the identity map. A function f will be called <u>hypoconvex</u> if $f \in C^*_\chi(X)$ for some hypoconvex set X. If Y is a hypoconvex subset in $\mathfrak{M}_n$ and range $f \subset Y$, we define $f_* : EN_n(X) \to EN_n(Y)$, as follows: given an essentially n-normal operator T in $\mathcal{L}(\mathcal{H})$ such that $R_e^n(T) = X$, let $\tau : C_X^*(X) \to \mathfrak{Q}(\mathcal{H})$ be the $*$-monomorphism determined by $\tau X = \pi T$. Also, let T_f be any operator in $\mathcal{L}(\mathcal{H})$ such that $\pi T_f = \tau f(X)$. (Notice that T_f is essentially n-normal and $R_e^n(T_f) \subset Y$). Now, we let $f_*[T] = RT_f \oplus S]$, where S is any n-normal operator in $\mathcal{L}(\mathcal{H})$ such that $R_e^n(S) = Y$. (Observe that f_* is well defined since $R_e^n(T_f \oplus S) = (R_e^n(T_f) \cup R_e^n(S))^\wedge = R_e^n(S) = Y$, see [16, Lemma 1.4]). Furthermore, if S can be represented by an $n \times n$ matrix of commuting normal operators S_{ij} whose joint essential spectrum coincides with Y, and T can be represented by an $n \times n$ matrix (T_{ij}) of essentially commuting, essentially normal operators, we define $f_* : \mathrm{Ext}(X) \to \mathrm{Ext}(Y)$ by $f_*[T_{ij}] = [(T_f)_{ij} \oplus S_{ij}]$.

From the definitions of the maps β and f_*, we deduce that the following diagram is commutative

$$\begin{array}{ccc} \mathrm{Ext}(X) & \xrightarrow{\ \beta\ } & EN_n(X) \\ \big\downarrow f_* & & \big\downarrow f_* \\ \mathrm{Ext}(Y) & \xrightarrow{\ \beta\ } & EN_n(Y) \end{array}$$

The following theorem implies that $EN_n(\cdot)$ is a homotopy invariant, covariant functor from the category of hypoconvex subsets of $\mathfrak{M}_n$ (with morphisms consisting of the class of hypoconvex functions) into the category of abelian groups.

THEOREM 2.9. [16, Theorem 3.17]. Let X and Y be two hypoconvex subsets of $\mathfrak{M}_n$ and let $f_t : X \to Y$, $0 \leq t \leq 1$, be a continuous arc of hypoconvex functions. Then $f(0)_* = f(1)_*$.

<u>Proof</u>. It is a direct consequence of the homotopy invariance of $\mathrm{Ext}(\cdot)$ (see [4, §2] and also [7]) and the commutativity of the following diagram for $t = 0$ and $t = 1$.

$$\begin{array}{ccccc} \mathrm{Ext}(X) & \xrightarrow{\beta} & EN_n(X) & \longrightarrow & 0 \\ {\scriptstyle (f_t)_*}\downarrow & & \downarrow{\scriptstyle (f_t)_*} & & \\ \mathrm{Ext}(Y) & \xrightarrow{\beta} & EN_n(Y) & \longrightarrow & 0 \end{array}$$

REMARK 2.10. From Theorem 2.9 we deduce that if B_n denotes the closed unit ball in $\mathfrak{M}_n$, then $EN_n(B_n) = \{0\}$. In particular, it follows that if T is an essentially n-normal operator such that $R_e^n(T) = B_n$, then T is a compact perturbation of an n-normal operator (see [16, Lemma 3.3]).

Definition 2.11.

Let X be a hypoconvex subset in $\mathfrak{M}_n$. We denote by X_N the normal part of X, and by $\widetilde{X}$ the quotient space of all unitary orbits of X.

THEOREM 2.12. [10, Theorem 3.6, Theorem 4.6], [16, Theorem 3.9]. Let X be a hypoconvex subset of $\mathfrak{M}_n$, and suppose that there exists a continuous cross section $s : \widetilde{X} \to X$ of the canonical quotient map from X onto $\widetilde{X}$.

(a) If X consists only of irreducible matrices, then $\beta \circ s_* : \mathrm{Ext}(\widetilde{X}) \to EN_n(X)$ is an isomorphism.

(b) Let $X \subset \mathfrak{M}_2$. If $\widetilde{X}_N$ is a strong deformation retract of $\widetilde{X}$, and $i : X_N \to X$ is the canonical inclusion, then $i_* : EN_2(X_N) \to EN_2(X)$ is an isomorphism. In particular, if $\beta \circ s_* : \mathrm{Ext}(\widetilde{X}_N) \to EN_2(X_N)$ is an isomorphism, then so is $\beta \circ s_* : \mathrm{Ext}(\widetilde{X}) \to EN_2(X)$.

REMARK 2.13. (a) In [9], a continuous section of the canonical map from X onto $\widetilde{X}$ was constructed, for every compact subset X of $\mathfrak{M}_2$ closed under unitary conjugation. On the other hand, it is also shown in [9] that if $n > 2$, there exists a hypoconvex subset $X \subset \mathfrak{M}_n$ such that there exists no such continuous section.

(b) Let $X = \left\{ \begin{pmatrix} \lambda & 1 \\ 0 & \bar{\lambda} \end{pmatrix} : |\lambda| = 1 \right\}^{\wedge}$ and let $Y = \left\{ \begin{pmatrix} \lambda & 1 \\ 0 & \bar{\lambda} \end{pmatrix} : |\lambda| = 1, \operatorname{Im} \lambda \geq 0 \right\}$. Then any two different matrices of Y are not equivalent and $\hat{Y} = X$. Thus, $p \mid Y : Y \to \widetilde{Y} = \widetilde{X}$ is a homeomorphism (that is, Y is a generating set for X, [16]) and since Y is a continuous arc, it follows from Theorem 2.12 (a) that

$EN_2(X) \simeq Ext(\tilde{X}) = \{0\}$. In particular, $\begin{pmatrix} U^+ & 1 \\ 0 & (U^+)^* \end{pmatrix}$ is weakly compalent to a binormal operator, [10].

(c) (See [10]). Let $X = \left\{\begin{pmatrix} \lambda & t \\ 0 & \lambda \end{pmatrix}, 0 \leq t \leq 1, |\lambda| = 1\right\}^\wedge$. Then it is easy to see that $\tilde{X}_N$ is homeomorphic to the Möbius band, and that $\tilde{X}$ consists of the Möbius band together with an annulus glued to the boundary of the Möbius band. Since $EN_2(X_N) \simeq Ext(S^1)$ ($= \mathbb{Z}$), we see from Theorem 2.12 (b) that $EN_2(X) \simeq \mathbb{Z}$. Furthermore, using the fact that $[U^+]$ is a generator for $EN_2(X_N)$, it follows that $[U^+ \oplus S]$ is a generator for $EN_2(X)$, where S is any 2-normal operator such that $R_e^2(S) = X$. Therefore, the mapping $[T] \to \mathrm{ind}(T)$ establishes an isomorphism between $EN_2(X)$ and $\mathbb{Z}$. In particular, $\begin{pmatrix} U^+ & 0 \\ 0 & U^+ \end{pmatrix} \oplus S \approx \begin{pmatrix} U^+ & 1 \\ 0 & U^+ \end{pmatrix} \oplus S$. (This fact can also be proved using the homotopy invariance result of [16, Theorem 4.8], see Remark 4.9.)

(d) In general, the conclusion of Theorem 2.12 is not valid (see [10, §4]). If $X \subset \mathfrak{M}_2$, and $j : \tilde{X} \to X$ is the continuous section constructed in [9] (see part (a) of this remark), then $\beta \circ j_*$ defines a map from $Ext(\tilde{X})$ into $EN_2(X)$. All one can hope is that this map is surjective (see [11, Remark 5.1]). Indeed, let $Y = (X \cup \{0\})^\wedge$, where X is as in part (c) of this remark. It is easy to see that $\tilde{Y}_N$ and $\tilde{Y}$ are the disjoint union of $\tilde{X}_N$ and $\tilde{X}$ with a homeomorphic copy of $S^1 \cup \{0\}$, respectively. Since $EN_2(Y_N) \simeq Ext(S^1 \cup \{0\}) \simeq \mathbb{Z}$, we deduce from Theorem 2.12 (b) that $EN_2(Y) = \mathbb{Z}$. But, $Ext(\tilde{Y}) = Ext(\tilde{X}) \oplus Ext(S^1 \cup \{0\}) \simeq \mathbb{Z} \oplus \mathbb{Z}$, and hence $\beta \circ j_*$ cannot be one-to-one, in this case.

(e) There is an alternative way of defining $EN_n(X)$. Following [2] and [4], given a unital separable C^*-algebra $\mathfrak{A}$, we shall denote by $Ext(\mathfrak{A})$ the set of weak equivalence classes of unital $*$-monomorphisms from $\mathfrak{A}$ into $\mathfrak{D}(\mathcal{H})$. Given a hypoconvex set $X \subset \mathfrak{M}_n$, let $Ext_X^n(X)$ be the subgroup of $Ext(C_X^*(X))$ consisting of those unital $*$-monomorphisms $\tau : C_X^*(X) \to \mathfrak{D}(\mathcal{H})$ such that $\tau x = \pi T$, where T is an essentially n-normal operator with $R_e^n(T) = X$. In [15] and [16] it is shown that $EN_n(X)$ is isomorphic to $Ext_X^n(X)$. Thus, information about the group $Ext(C_X^*(X))$, together with knowledge about $Ext(\tilde{X})$, seem to be relevant for the classification problem of essentially n-normal operators, up to weak compalence.

3. TWO EXAMPLES OF NON-TRIVIAL ESSENTIALLY n-NORMAL OPERATORS.

In this section, we present two interesting examples of essentially n-normal operators, whose properties were discussed in detail in [11]. The first example shows in a very explicit way a torsion phenomenon already discovered in [6]. The second provides us with an example of a hypoconvex set $X \subset \mathfrak{m}_2$ such that $\mathrm{Ext}^2_X(X)$ is a proper subgroup of $\mathrm{Ext}(C^*_X(X))$. Also, this operator answers the question raised in [14], concerning algebraically essentially n-normal operators.

Example 3.1.

Let M be a normal operator in $\mathcal{L}(\mathcal{H})$ whose spectrum coincides with the unit disk, and let $S = U^+ \oplus \begin{pmatrix} M & 1-|M| \\ 0 & M \end{pmatrix}$, where U^+ denotes, as usual, a simple unilateral shift. It readily follows that $R^2_e(S) = \left\{ \begin{pmatrix} \lambda & 1-|\lambda| \\ 0 & \lambda \end{pmatrix}, \ |\lambda| \le 1 \right\}^{\wedge}$. Let $Y = R^2_e(S)$. Then a moment's reflection show that $\tilde{Y}_N$ is homeomorphic to a Möbius band, and $\tilde{Y}$ is homeomorphic to a Möbius band with a disk glued to its boundary. Therefore, $\tilde{Y}$ is homeomorphic to the real projective plane.

Given a hypoconvex set $X \subset \mathfrak{m}_n$, let $\mathfrak{J}_X$ denote the ideal in $C^*_X(X)$ consisting of all those functions that vanish on X_N. Then, $C^*_X(X)/\mathfrak{J}_X$ is *-isomorphic with $C^*_X(X_N)$, and the sequence

$$0 \to \mathfrak{J}_X \to C^*_X(X) \to C^*_X(X_N) \to 0$$

is exact. It follows from [2, §6] that the sequence

$$(*) \qquad \mathrm{Ext}(C^*_X(X_N)) \to \mathrm{Ext}(C^*_X(X)) \to \mathrm{Ext}(\mathfrak{J}_X)$$

is also exact.

In our situation, $C^*_X(Y_N)$ is *-isomorphic with $C(S^1)$, and $\mathfrak{J}_Y$ is *-isomorphic with the ideal of $C(S^2) \otimes \mathfrak{m}_2$ consisting of those functions that vanish at the north pole. From [3, Theorem 3.1] and [6], it follows that $\mathrm{Ext}[C^*_X(Y_N)] \simeq \mathrm{Ext}(S^1) \simeq \mathbb{Z}$, and $\mathrm{Ext}[\mathfrak{J}_Y] \simeq \mathrm{Ext}(S^2) \simeq \{0\}$. Using the exactness of the sequence (*) for the case $X = Y$, we deduce that $\mathrm{Ext}[C^*_X(Y)]$ is a cyclic group, and a generator can be obtained by calculating the image of the generator of $\mathrm{Ext}(S^1)$. From [3, Theorem 3.1], we know that a generator of $\mathrm{Ext}(S^1)$ is induced by the *-monomorphism from $C(S^1)$ into $\mathfrak{Q}(\mathcal{H})$ whose value at the coordinate function of

S^1 is $\pi(U^+)$. Since the trivial extension of $C^*_X(Y)$ is defined by sending X to $\begin{pmatrix} M & 1-|M| \\ 0 & M \end{pmatrix}$, a generator for $\mathrm{Ext}(C^*_X(Y))$ is uniquely determined by the $*$-monomorphism from $C^*_X(Y)$ into $\mathfrak{D}(\mathcal{H})$ whose value at X is $\pi\left(U^+ \oplus \begin{pmatrix} M & 1-|M| \\ 0 & M \end{pmatrix}\right) = \pi(S)$. In particular, $EN_2(Y) \simeq \mathrm{Ext}(C^*_X(Y))$. Employing the fact that the C^* double suspension of $\mathfrak{M}_2$, $S^2(\mathfrak{M}_2)$ is $*$-isomorphic with $C^*_X(Y)$, and that $\mathrm{Ext}(S^2(\mathfrak{M}_2)) = \mathbb{Z}_2$ (see [11, §3]), we conclude that $EN_2(Y) = \mathbb{Z}_2$. Furthermore, we also deduce that $[S] \neq 0$, while $[S] + [S] = 0$. This means that S is not weakly compalent to a 2-normal operator, while $S \oplus S$ is. The fact that $[S \oplus S] = 0$ can be proved with a simpler and direct argument. Indeed, notice that

$$(**) \qquad S \oplus S \approx \begin{pmatrix} U^+ \oplus M & 0 \oplus (1-|M|) \\ 0 & U^+ \oplus M \end{pmatrix}.$$

Since $0 \oplus (1-|M|) = 1 - |U^+ \oplus M|$, and $U^+ \oplus M \approx M$, each of the entries of the right hand side of (**) are compact perturbations of elements in $C^*(M)$, as desired.

Example 3.2.

Let T be the 4×4 matrix of essentially commuting essentially normal operators given by

$$T = \begin{pmatrix} 0 & 1 & -i\sqrt{1-\mathrm{Re}\,U^+} & 0 \\ U^+ & 0 & \frac{1}{\sqrt{2}}(1-U^+) & 0 \\ 0 & 0 & -1 & 0 \\ i\,U^+\sqrt{1-\mathrm{Re}\,U^+} & \frac{1}{\sqrt{2}}(1-U^+) & 0 & 1 \end{pmatrix}.$$

An easy verification shows that T is a Fredholm operator. To compute its index, we notice that T can be regarded as a 2×2 matrix $\begin{pmatrix} A & B \\ C & D \end{pmatrix}$ of essentially 2-normal operators, and that $D = \begin{pmatrix} -1 & 0 \\ 0 & 1 \end{pmatrix}$ is invertible. Therefore $\mathrm{ind}(T) = \mathrm{Ind}(A - BD^{-1}C) = \mathrm{Ind}(U^+) = -1$. Now, we show that T is algebraically essentially 2-normal ([15], [14]), that is, that every irreducible representation of $\pi(C^*(T))$ is at most 2-dimensional. To see this, we observe that $\pi C^*(T)$ is $*$-isomorphic with the C^*-subalgebra of $C[0,1] \otimes \mathfrak{M}_4$ generated by the matrix-valued function

$$A(t) = \begin{pmatrix} 0 & 1 & g(t) & 0 \\ e^{2\pi it} & 0 & \frac{1}{\sqrt{2}}\, h(t) & 0 \\ 0 & 0 & -1 & 0 \\ -e^{2\pi it} g(t) & \frac{1}{\sqrt{2}}\, h(t) & 0 & 1 \end{pmatrix},$$

where $g(t) = -i\,(1-\cos 2\pi t)^{1/2}$, and $h(t) = 1 - e^{2\pi it}$, $0 \le t \le 1$. If we set

$$U(t) = \begin{pmatrix} \frac{1}{\sqrt{2}}\, e^{\pi it} & \frac{1}{\sqrt{2}} & 0 & 0 \\ 0 & 0 & 1 & 0 \\ 0 & 0 & 0 & 1 \\ -\frac{1}{\sqrt{2}}\, e^{\pi it} & \frac{1}{\sqrt{2}} & 0 & 0 \end{pmatrix}.$$

then U is a unitary element of $C[0,1] \otimes \mathfrak{m}_4$, and

$$U(t)A(t)U^*(t) = \begin{pmatrix} e^{\pi it} & h(t) & 0 & 0 \\ 0 & -1 & 0 & 0 \\ 0 & 0 & 1 & h(t) \\ 0 & 0 & 0 & -e^{\pi it} \end{pmatrix}.$$

The C^*-algebra generated by the above matrix-valued function is clearly $*$-isomorphic to the C^*-subalgebra of $C[-1,1] \otimes \mathfrak{m}_2$ generated by the matrix-valued function

$$B(t) = \begin{cases} \begin{pmatrix} e^{-\pi it} & h(-t) \\ 0 & -1 \end{pmatrix}, & \text{for } -1 \le t \le 0 \\ \begin{pmatrix} 1 & h(t) \\ 0 & -e^{\pi it} \end{pmatrix}, & \text{for } 0 \le t \le 1. \end{cases}$$

It is clear that the C^*-algebra generated by B has no irreducible representation of dimension greater than 2 and thus we see that T is an essentially algebraically 2-normal operator. Let $X = R_e^2(T)$ $(= (\text{range } B)^\wedge)$. Since two different matrices in range (B) are not unitarily equivalent, it follows that $\tilde{X}$ is homeomorphic to $[-1,1]$ via the map $t \to \tilde{B}(t)$. Therefore, $\mathrm{Ext}(\tilde{X}) = \{0\}$. Indeed, using arguments involving the Mayer-Vietoris sequence and exactness of $\mathrm{Ext}(\cdot)$, it can be shown that $\mathrm{Ext}(X) = \{0\}$ (see [11, §5]). Thus, employing Theorem 2.6, we conclude that $EN_2(X) = \{0\}$. In particular, every essentially 2-normal operator R such that $R_e^2(R) = X$ must be a compact perturbation of a 2-normal operator, and hence, must have index zero. Since $\mathrm{ind}\, T \neq 0$, we see that T cannot be essentially 2-normal. Furthermore, it readily follows that T cannot be the direct sum of an essentially 2-normal operator and an essentially normal operator. This answers negatively a question raised in [14]. (Recall [8] that if R is algebraically n-normal, then $R = \sum_{k=1}^{n} \oplus R_k$, where R_k is k-normal, $1 \leq k \leq n$, and some of the R_k's might be missing.) On the other hand, if τ is the *-monomorphism from $C_X^*(X)$ into $\mathfrak{Q}(\mathcal{H})$ determined by $\tau X = \pi T$, we see that τ is non-trivial and therefore $\mathrm{Ext}(C_X^*(X)) \neq \{0\}$. Thus, in this case, $\mathrm{Ext}_X^2(X)$ $(=\{0\})$ is a proper subgroup of $\mathrm{Ext}(C_X^*(X))$, as desired.

A reasonable program for classifying essentially n-normal operators is to try to calculate the group $EN_n(X)$ for every hypoconvex set $X \subset \mathfrak{M}_n$. The examples presented in this section reveal some of the subtleties involved in this problem.

REFERENCES

[1] W. Arveson, A note on essentially normal operators, Proc. Roy. Irish Acad., 74 (1974), 143-146.

[2] L. G. Brown, Extensions and the structure of C*-algebras, Symp. Mat., XX, 539-566, Academic Press, 1976.

[3] L. G. Brown, R. G. Douglas and P. A. Fillmore, Unitary equivalence modulo the compact operators and extensions of C*-algebras, 58-128, Proceedings of a Conference in Operator Theory, Lecture Notes in Mathematics, No. 345, Springer-Verlag, New York, 1973.

[4] L. G. Brown, R. G. Douglas and P. A. Fillmore, Extensions of C*-algebras and K-homology, Ann. of Math., 105 (1977), 265-324.

[5] J. W. Bunce and J. A. Deddens, Irreducible representations of the C*-algebra generated by an n-normal operator, Trans. Amer. Math. Soc., 171 (1972), 301-307.

[6] R. G. Douglas, The relation of Ext to K-theory, Symp. Math., XX, 513-531, Academic Press, 1976.

[7] D. P. O'Donovan, Quasidiagonality in the Brown-Douglas-Fillmore theory, Duke Math. J., 44 (1977), 767-776.

[8] C. Pearcy, A complete set of unitary invariants for operators generating finite W^*-algebras of type I, Pacific J. Math., 12 (1962), 1405-1416.

[9] V. Paulsen, Continuous canonical forms for matrices under unitary equivalence, Pacific J. Math., (to appear).

[10] V. Paulsen, Weak compalence invariants for essentially n-normal operators. Preprints.

[11] V. Paulsen and N. Salinas, Two examples of non-trivial essentially n-normal operators. Preprint.

[12] C. Pearcy and N. Salinas, Finite dimensional representations of C^*-algebras and the matricial reducing spectra of an operator, Rev. Roum. Math. Pures et Appl. (Bucarest), 20, (1975), 567-598.

[13] C. Pearcy and N. Salinas, The reducing essential matricial spectra of an operator, Duke Math. J., 42 (1975), 423-434.

[14] C. Pearcy and N. Salinas, Extensions of C^*-algebras and the reducing essential spectra of an operator. K-theory and operator algebras, Springer-Verlag Lecture Notes in Mathematics, No. 575, 96-112.

[15] N. Salinas, Extensions of C^*-algebras and essentially n-normal operators, Bull. Amer. Math. Soc., 82 (1976), 143-146.

[16] N. Salinas, Hypoconvexity and essentially n-normal operators, Trans. Amer. Math. Soc. To appear.

[17] N. Salinas, Homotopy invariance of Ext($\mathfrak{a}$), Duke Math. J., 44 (1977), 777-794.

THE UNIVERSITY OF KANSAS
LAWRENCE, KANSAS 66045

SOME PROBLEMS IN OPERATOR THEORY

Allen L. Shields

Notes by Michael J. Hoffman*

I'm going to discuss a few problems in operator theory. Some of them I had something to do with; some I didn't. I won't prove anything, but will try to indicate some areas of interest. Unless specified otherwise, these will refer to bounded linear operators on Hilbert space.

1. Hyperinvariant subspaces. One such area is the continuing interest in extensions of the ideas of Victor Lomonosov. He proved that if you have an arbitrary operator A which commutes with a nonzero compact operator, then you can conclude that A has a nontrivial hyperinvariant subspace, [10], [12]. In this discussion subspace will always mean closed subspace. Suppose we have a subspace M of a Hilbert space H and a bounded linear transformation, that is an operator, A acting on H. Ordinary invariance says that M as a set is mapped into itself. $AM \subseteq M$. Such a subspace is called invariant. It is called hyperinvariant if it is mapped into itself by all operators that commute with A. That is, if for every T such that $AT = TA$ we have $TM \subseteq M$, we say M is a hyperinvariant subspace for A.

It is still an open problem whether every operator has a nontrivial invariant subspace. In the case of a compact operator, Aronszajn and Smith had proved, and evidently von Neumann had earlier proved, that a compact operator has an invariant subspace [1]. But it was an open question whether two commuting compact operators have a common invariant subspace. That remained open until 1973 when Lomonosov proved that in fact a nonzero compact operator has a hyperinvariant subspace. So you actually get an invariant subspace common to much more than just a pair of commuting compact operators.

---

*The notes for this talk were written, revised, and references added, by Michael Hoffman. The references are intended as starting points for interested readers and are not intended to be exhaustive. Apologies are in order to those whose work has not been mentioned.

THEOREM (Lomonosov). If A is a non-scalar operator on H and there is a nonzero compact operator K such that $AK - KA = 0$, then A has a hyperinvariant subspace.

A few years ago, John Daughtry had a nice idea for extending this result. Instead of 0, you should put a small operator R on the right side of the equation in Lomonosov's theorem. $AK - KA = R$. In what sense should R be small? Of course compactness is one kind of "small," but that would be meaningless here as the left side is automatically compact. Small norm wouldn't do anything because we are only interested in the existence of hyperinvariant subspaces and could assume that A had small norm to begin with. Daughtry had the idea of letting R be an operator of rank 1. He was able to show that this implies that A has an invariant subspace [3]. Then H. W. Kim, Carl Pearcy, and I modified his argument a little bit to show that you could actually get hyper-invariant subspaces out of this [8].

THEOREM (Daughtry). If $AK - KA = R$ where K is nonzero and compact and R has rank 1, then A has a hyperinvariant subspace.

This suggests at least a couple of questions. One is, "why rank one." What about rank two? Of course, there is an immediate triviality. If you allowed the compact operator K to have rank 1, then $AK - KA$ would have rank at most 2. So you can attain rank 2 trivially, but that is sort of cheating. Just as in Lomonosov's result K must not be the zero operator, so here you would want the rank of K to be greater than 1. But it turns out that doesn't help. If A is any operator that is not a scalar multiple of the identity, you can find a K which has rank 10, if you like, such that $AK - KA$ has rank 2. See [9, Prop. 4]. So the rank 2 hypothesis doesn't seem to have any force to it. But rank 1 does, and you get the existence of hyperinvariant subspaces. The direction of further extending the right hand side doesn't seem to lead anywhere.

But now there is another question. Which operators satisfy the conditions of the theorem? We introduce a notation. Let $\Delta(H)$ denote the set of operators A for which there exists a nonzero compact operator K such that the rank of $AK - KA$ is less than or equal to 1. The first surprising thing is that I don't know any operators which are not in this class. I will state the first question in a bold way.

Question 1. Is every operator in $\Delta(H)$?

Presumably not, but find one. It is surprising that $\Delta(H)$ does contain many operators for which it would not immediately be obvious. For example, the unilateral shift does not commute with any nonzero compact operators. On the other hand, its adjoint has an eigenvector, and one shows rather easily that any operator such that either it or its adjoint has an eigenvector is in this class [9, Prop. 2]. That takes care of the shift. The class also contains any operator with disconnected spectrum, [9, Th. 3], and any non-scalar normal operator, [9, Cor. 7]. For example, multiplication by x on L^2 of the unit interval does not commute with any nonzero compact operator. So it does not satisfy the conditions for Lomonosov's theorem. But it is in this class $\Delta(H)$. There is a rank one commutator with some compact operator. In fact, n-normal operators are in $\Delta(H)$, [9, Cor. 7].

I'll mention a question suggested by the n-normal situation and then leave this area. An operator is called n-normal if it can be represented as an $n \times n$ matrix whose entries are commuting normal operators. But it is known [4] that by a unitary equivalence you can put it into the form of an upper triangular matrix. So we'll assume that has been done and our n-normal operator already has the form

$$A = \begin{bmatrix} A_{11} & A_{12} & A_{13} & \cdots & A_{1n} \\ 0 & A_{22} & A_{23} & \cdots & A_{2n} \\ 0 & 0 & A_{33} & \cdots & A_{3n} \\ & 0 & & & \vdots \\ & & & & A_{nn} \end{bmatrix}$$

where the entries A_{ij} are commuting normal operators, and A operates on the direct sum of n copies of the Hilbert space. This is in $\Delta(H)$, but, when you examine the proof, you find you have used almost nothing. You use only that it is in the upper triangular form and that the two extreme entries, A_{11} in the upper left corner and A_{nn} in the lower right, are normal; not even that they commute. So you actually get that if A has the above form with A_{11} and A_{nn} normal, then A has hyperinvariant subspaces. In fact A is in $\Delta(H)$. But you do use the fact that there is not only an upper left entry which is normal, but also one on the lower right. That suggests the following question which I don't see how to do anything with. Suppose instead of n copies of the Hilbert space, we take countably infinitely many. Again consider an upper triangular matrix, but now, of course, with no lower right corner.

$$A = \begin{bmatrix} A_{11} & A_{12} & A_{13} & \cdots \\ 0 & A_{22} & A_{23} & \cdots \\ 0 & 0 & A_{33} & \cdots \\ \cdot & \cdot & & \\ \cdot & \cdot & & \end{bmatrix} .$$

Again we start with the A_{ij} as commuting normal operators. Of course, it is not automatic that if you just write down something like this that it is going to be an operator. We must assume it acts as a bounded operator on $H \oplus H \oplus H \oplus \cdots$.

Question 2. If A can be represented as above, is A in $\Delta(H)$?

If A is not a scalar, then does A have a hyperinvariant subspace? The argument which worked before does not work. It depended on having a stopping point in the lower right corner.

Having posed a question which I don't see how to answer, let me go beyond that and make it perhaps more difficult. In the n-normal case we observed that we didn't really need all the entries to be normal, just the first and last on the main diagonal. Suppose we did the same thing here. There is no last, so we only assume A_{11} to be normal. Then we would really be looking at an operator with an invariant subspace, the first component, such that the restriction to that invariant subspace is a normal operator.

Question 3. Suppose A is an operator with an invariant subspace M such that $A|_M$ is normal. So we are talking about an extension of a normal operator.

Does A have a hyperinvariant subspace? Is A in $\Delta(H)$?

I learned just yesterday of some work done by Jaime Bravo, a graduate student at Berkeley. He has proved some interesting things. One I want to mention is a question he has raised which is somewhat in this general direction.

Question 4. Suppose A is an operator and P is a projection of rank 1 such that $(AP - PA)^3 = 0$. Does this imply that A has a hyperinvariant subspace?

One might have many reactions to this question ranging from "so what?" to "why cubed?" Let me look at the second of these. If you took the first power, $AP - PA = 0$, then with any proper projection you have a reducing subspace. If the rank of P is 1, you have an eigenvector. The resulting eigenspace is certainly hyperinvariant. Implicit in this is that A is not a scalar multiple of the identity. What about the second power? It is a little bit of work then, which I won't go through, but if $(AP - PA)^2 = 0$ you still get a one-dimensional invariant subspace. So A has an eigenvector, and the eigenspace for the corresponding eigenvalue is hyperinvariant. Thus the cube is the first power for which you don't seem to get a result for elementary reasons.

One might still ask, "So what?" Granted the cube is the right power to begin with, who cares? This is just one more condition which would be nice if you could prove it. But this one would have several immediate applications if one could do it. First of all, every bilateral weighted shift of multiplicity one does satisfy this. There is such a P, namely projection onto one of the coordinates, say the 0th coordinate. It is not known whether every bilateral shift has a hyperinvariant subspace. It is known that they have invariant subspaces. It is known that if a bilateral shift is not invertible, then it has a hyperinvariant subspace. In fact every invariant subspace is hyperinvariant. See [5, Th. 1 and Cor. 2]. But for invertible bilateral shifts it is not known whether or not hyperinvariant subspaces exist. For some classes it is known, but not in general. For some information see the paper of Ralph Gellar and Domingo Herrero [5] and Section 10, p. 109 and Section 12, p. 119 of the survey paper [15].

Question 5. Does every bilateral weighted shift have a hyperinvariant subspace?

An affirmative answer to Question 4 would give an affirmative answer to Question 5.

Another question which has been around for a long time which I think was first posed by Ronald Douglas is the following.

Question 6. Suppose you have an operator A which has an invariant subspace and is invertible. Does it follow that A^{-1} has an invariant subspace?

If an affirmative answer to Question 4 could be proved, then using a theorem Bravo has established, you could answer the question of Douglas affirmatively. The answer to Question 6 is yes if that to Question 4 is yes. So this certainly seems to be a conjecture worth looking at since it has these nice applications and is intrinsically interesting. That is as much as I want to say about invariant subspaces.

2. Composition operators. I want to mention a couple of questions that have to do with composition operators. These were posed by Raymond Roan who was a student at Michigan and is now at Kentucky. He considers composition operators not on the space H^p but rather on the space of analytic functions on the unit disc whose first derivatives lie in H^p. Let $1 \leq p < \infty$, and put $D = \{z \in \mathbf{C} : |z| < 1\}$. Let $S^p = \{f : f$ is analytic on D and f' is in $H^p\}$. Roan asks, "What are the composition operators on S^p?" The norm on S^p is essentially the H^p - norm of the derivative, $\|f'\|_p^p = (1/2\pi) \int_0^{2\pi} |f'(t)|^p \, dt$. This isn't quite a norm because constant functions would have norm 0. You can take care of that by saying, for example, that the norm be given by $\|f\| = |f(0)| + \|f'\|_p$. This particular norm is, in a sense, a bit unnatural. Why take the sum of the two instead of the maximum or the square root of the sum of the squares? Or why $f(0)$? But you do something to take care of the constants and to take into account the H^p norm of the derivative.

One only talks about composition operators, of course, for functions that map the disc into itself. So we assume φ is analytic in D and maps D into D. Let C_φ be the composition operator obtained by composing with φ; $C_\varphi(f) = f \circ \varphi$. The question is, when does this operate on the space? By an easy application of the closed graph theorem, if it operates in the sense that every function in the space is mapped to another function in the space, then it must do so in a bounded manner.

Question 7. For which φ does C_φ operate on S^p?

This is not known. A necessary condition is that the function φ must itself be in the class you are talking about. C_φ operates on S^p only if $\varphi \in S^p$. But this is not sufficient. Roan does have some results, which I won't write out in detail, giving sufficient conditions for something to operate. But the general question is open. Also open is:

Question 8. When is C_φ compact?

The results seem to be rather different from the H^p situation and yet there is still some connection between the two. It looks like an area that might be interesting. See [13] and [14].

3. Compact approximants. Finally I will mention one problem, or rather two problems of the same general sort, which were suggested by some work that Sheldon Axler and I were doing. It is known that if you have an operator on Hilbert space then you can always find a nearest compact operator. There is at least one compact operator which minimizes the distance to the set of all compact operators. This is proved by Richard Holmes and Bernard Kripke in [7].

THEOREM (Holmes, Kripke). If A is an operator on Hilbert space, there is a nearest compact operator.

There is no trivial reason why there should exist a nearest one because the compact operators aren't "compact." That is, the unit ball of the compact operators is not a compact set in any reasonable sense. Nonetheless, there do exist nearest ones. One general type of question concerns special classes of operators. We would like the nearest compact operator to again lie in the special class. Or at any rate, among the many possible nearest ones, there is no reason for it to be unique, at least one lies in the class. The special class in which we are interested is the set of Hankel operators. We fix an orthonormal basis $(e_n)_{n=1}^\infty$ and speak of Hankel operators as those for which the matrix is constant along the diagonals perpendicular to the main diagonal. We have a sequence $a_1, a_2, a_3, \ldots$ along the first row. The second row is shifted over one, beginning with a_2. The third row begins $a_3, a_4, a_5, \ldots$ and so on.

$$A = \begin{bmatrix} a_1 & a_2 & a_3 & a_4 & \cdots \\ a_2 & a_3 & a_4 & a_5 & \cdots \\ a_3 & a_4 & a_5 & a_6 & \cdots \\ \vdots & \vdots & \vdots & \vdots & \end{bmatrix}$$

That's a Hankel operator. Of course, saying that it is an operator assumes that it is a bounded operator on ℓ^2. It is known when this is bounded and when it is compact [11], [6].

Question 9. If A is a Hankel operator, does there exist a nearest compact operator which is also Hankel?

(This question has been settled. See below.)

There is a reason for posing this question. In function theory one is sometimes interested in the existence of nearest elements. For instance, it is known that if I have an L^∞-function f on the circle, then there is a nearest element in H^∞. But suppose one asks about $H^\infty + C$ where C represents the space of continuous functions on the unit circle. Is there a nearest element to f in $H^\infty + C$? That is not known. An affirmative answer to the operator problem would imply existence in the approximation problem.

Finally, the question of nearest compact operators raises another question with which I'll leave you. If X is a Banach space, let $K(X)$ denote the set of compact operators on X. Suppose T is any operator on X and look at the distance from T to $K(X)$. Now do the same thing in the dual space. Look at the distance from the adjoint T^* to the set of compact operators on the dual space. Are they the same?

Question 10. Do we have $\mathrm{dist}(T, K(X)) = \mathrm{dist}(T^*, K(X^*))$?

We can prove this with some additional conditions, but not in general.

(Work on this question has also progressed. See below.)

Thank you.

Since this talk was given Question 9 has been answered affirmatively. This done in the preprint, "Approximation by Compact Operators and the Space $H^\infty + C$" by Sheldon Axler, I. David Berg, Nicholas Jewell, and Allen Shields. This paper also contains material relating to question 10. They establish inequalities under additional assumptions and give an example of a Banach space operator for which the essential norm of the adjoint is one half the essential norm of the operator.

Questions from the audience.

Question: Suppose in the setting of Daughtry's theorem that instead of having rank 1 you require that R be small in norm by comparison to A and K. For example, suppose $\|A\| \geq 1$ and there is a non-zero compact operator K such that $\|K\| \geq 1$ and $\|AK - KA\| < 1/2$. Could you do anything then? Or might the situation be like that for rank 2, that you can always make R small this way?

Response: I don't know. Try it.

Question: To show that n-normal operators are in $\Delta(H)$, you used the existence of a stopping point in the lower right corner. In Question 2 could you get anywhere with an assumption that the A_{nn} are getting small in some appropriate sense as you work your way down the diagonal?

Response: Again, I don't know. Try it.

Question: Can you relate Question 3 to the general hyperinvariant subspace problem by starting with any operator and considering a direct sum of it with a normal operator? This has a normal restriction to an invariant subspace.

Response: You would need to know that the hyperinvariant subspace you get for the direct sum splits properly and that it gives you a non-trivial (i.e., not 0 and not the whole space) hyperinvariant subspace for the operator you started with.

Question: Under what conditions have you been able to establish equality in Question 10?

Response: Equality holds if X is a conjugate Banach space. So, in particular, it is true for any reflexive Banach space. It is true a bit more generally. If X is a Banach space such that when you embed it canonically in its second dual there is a projection of norm 1 from the second dual onto the image of X, then the equation holds. This is true for L^1, for example, even though L^1 is not a conjugate Banach space. We may have some more technical conditions under which one can prove this even without such an assumption, but at least those were the easy cases when it was true.

Question: If you have an operator with an invariant subspace, what do you gain from hyperinvariance? Why work for that condition?

Response: You might take your question back one step and ask why try to get an invariant subspace. Why so much attention to that problem? It is partly, of course, that it is there. But also the hope is that, just as in the finite dimensional theory, by getting not just one invariant subspace, but enough of them, you might get simpler operators by restriction to these invariant subspaces. You might then begin to have some theory in which an operator could be decomposed into simpler pieces and you could really study them. For general operators that is still just a distant dream. Hyperinvariant subspaces would preseumably give you a bit more to work with in that situation.

Question: Is it perhaps true that a non-scalar operator is in $\Delta(H)$ if and only if it has a non-trivial hyperinvariant subspace?

Response: We have begun to suspect that this might be true and have tried to prove it, but haven't had any success yet. One thing is disappointing about everything we have been able to do with the class $\Delta(H)$. All the operators to which we could apply it, that is to say, all the operators we could prove lie in the class, were operators for which it was already known that there is a hyperinvariant subspace. In fact, the methods we had to use to get them into the class $\Delta(H)$ were very similar to those you use to prove the existence of a hyperinvariant subspace directly. So it kind of seems that something is going on in parallel to the general theory of hyperinvariant subspaces. Perhaps the conditions are, in fact, equivalent

REFERENCES

1. N. Aronszajn and K. T. Smith, Invariant subspaces of completely continuous operators, Annals of Math. 60 (1954), 345-350.

2. S. Axler, I. D. Berg, N. Jewell, and A. L. Shields, Approximation by compact operators and the space $H^\infty + C$, (preprint).

3. J. Daughtry, An invariant subspace theorem, Proc. Amer. Math. Soc., 49 (1975), 267-268.

4. D. Deckard and C. Pearcy, On matrices over the ring of continuous complex valued functions on a Stonian space, Proc. Amer. Math. Soc., 14 (1963), 322-328.

5. R. Gellar and D. A. Herrero, Hyperinvariant subspaces of bilateral weighted shifts, Indian Univ. Math. J., 23 (1974), 771-790.

6. P. Hartman, On completely continuous Hankel matrices, Proc. Amer. Math. Soc., 9 (1958), 862-866.

7. R. B. Holmes and B. R. Kripke, Best approximation by compact operators, Indiana Univ. Math. J., 21 (1971), 255-263.

8. H. W. Kim, C. Pearcy, and A. L. Shields, Rank-one commutators and hyperinvariant subspaces, Michigan Math. J., 22 (1975), 193-194.

9. H. W. Kim, C. Pearcy, and A. L. Shields, Sufficient conditions for rank-one commutators and hyperinvariant subspaces, Michigan Math. J., 23 (1976), 235-243.

10. V. Lomonosov, Invariant subspaces for operators commuting with compact operators, Funkcional Anal. i Prilozen, 7 (1973), 55-56 (Russian). Invariant subspaces for the family of operators which commute with a completely continuous operator, Functional Anal. and Appl., 7 (1973), 213-214 (English).

11. Z. Nehari, On bounded bilinear forms, Annals of Math., 65 (1957), 153-162.

12. C. Pearcy and A. L. Shields, A survey of the Lomonosov technique in the theory of invariant subspaces, Topics in Operator Theory, 219-229, Amer. Math. Soc. Surveys No. 13, Providence, R. I., 1974.

13. R. C. Roan, Composition operators on H^p with dense range, Indiana Univ. Math. J., 27 (1978), 159-162.

14. R. C. Roan, Composition operators on the space of functions with H^p derivative, (submitted).

15. A. L. Shields, Weighted shift operators and analytic function theory, Topics in Operator Theory, 49-128, Amer. Math. Soc. Surveys No. 13, Providence, R. I., 1974.

16. H. Widom, Hankel matrices, Trans. Amer. Math. Soc., 121 (1966), 1-35.

UNIVERSITY OF MICHIGAN

UNIVERSITY OF CALIFORNIA AT BERKELEY

ON A QUESTION OF DEDDENS

Joseph G. Stampfli*

Let $\mathcal{L}(\mathcal{H})$ denote the algebra of all bounded linear operators on a Hilbert space $\mathcal{H}$. For a fixed invertible operator A in $\mathcal{L}(\mathcal{H})$ set

$$\mathcal{B}_A = \{T \in \mathcal{L}(\mathcal{H}) : \sup_k \|A^k T A^{-k}\| \leq M_T < \infty \quad \text{for} \quad k = 1,2,\ldots\}$$

The algebra $\mathcal{B}_A$ seems to have been first introduced in [2]. During this conference (see [1]), James Deddens asked whether $\mathcal{B}_A = \mathcal{L}(\mathcal{H})$ implies that A is similar to a scalar multiple of a unitary. We shall show that indeed it does.

If $\lambda \in \partial\sigma_e(A)$, the boundary of the essential spectrum of A, then $\lambda \in \sigma_{\ell e}(A)$, the left essential spectrum of A and hence there exists an orthonormal sequence $\{f_k\}$ in $\mathcal{H}$ such that $\|(A - \lambda)f_k\| \to 0$. (See [3] for example.) We will have occasion to use this result in the proof. I am grateful to Donald Hadwin for pointing out an error in an earlier version of this paper.

THEOREM. Let A be an invertible operator in $\mathcal{L}(\mathcal{H})$. Then $\mathcal{B}_A = \mathcal{L}(\mathcal{H})$ if and only if A is similar to a scalar multiple of a unitary operator.

Proof. Since the sufficiency is obvious we turn to the necessity. First we multiply A by a non-zero scalar in order to assure that $1 \in \partial\sigma(A)$. We assume this has been done and denote the new operator by A as before.

Our immediate goal is to show that A is power bounded. Assume to the contrary that it is not, we proceed inductively as follows. We select an integer n_1 and a unit vector f_1 such that

$$\|A^{n_1} f_1\| \geq 1 .$$

Having selected integers $n_1,\ldots,n_k$ and orthonormal vectors $f_1,\ldots,f_k$ such that

$$\|A^{n_j} f_j\| \geq j \quad \text{for} \quad j = 1,2,\ldots,k$$

*The author gratefully acknowledges the support of the National Science Foundation.

we choose (<u>if possible</u>) an integer n_{k+1} and a unit vector f_{k+1} such that $f_{k+1} \perp M_k = \mathrm{sp}\,\{f_1,\ldots,f_k\}$ and $\|A^{n_{k+1}} f_{k+1}\| \geq k+1$. We distinguish two cases.

<u>Case</u> 1. The choice of the n_k's and f_k's may be continued indefinitely to yield an orthonormal sequence $\{f_k\}$ where $\|A^{n_k} f_k\| \geq k$ for all k. Next we observe that since $1 \in \partial\sigma_e(A)$, it follows that $1 \in \sigma_{\ell e}(A^{-1})$. Thus for each n_k we can choose a unit vector h_k such that

$$A^{-n_k} h_k = h_k + e_k$$

where $\|e_k\| \leq \|A^{n_k}\|^{-1}$ and moreover the h_k's are orthogonal. We now define a partial isometry V as follows

$$\begin{aligned} &V : h_k \to f_k \\ &V : g \to 0 \quad \text{if} \quad g \perp h_k \quad \text{for} \quad k = 1,2,\ldots\,. \end{aligned}$$

Thus

$$\begin{aligned} \|A^{n_k} V A^{-n_k}\| &\geq \|A^{n_k} V A^{-n_k} h_k\| = \|A^{n_k}(f_k + Ve_k)\| \\ &\geq \|A^{n_k} f_k\| - \|A^{n_k} Ve_k\| \geq k-1\,. \end{aligned}$$

Since the right hand side tends to ∞ with k this contradicts the hypotheses.

<u>Case</u> 2. The process of choosing the f_k's terminates. Thus there exists a finite dimensional subspace M and a constant C such that

$$\|A^n \mid M^\perp\| \leq C \quad \text{for} \quad n = 1,2,\ldots\,.$$

Hence $\|A^n \mid M\|$ must be unbounded and thus there exists a vector $f \in M$ such that $\|A^n f\|$ is unbounded. Choose a subsequence m_k such that $\|A^{m_k} f\| \geq k^2$ for $k = 1,2,\ldots$. Next choose an orthonormal sequence $\{h_k\}$ where

$$A^{-m_k} h_k = h_k + e_k \quad \text{and} \quad \|e_k\| \leq \|A^{m_k}\|^{-1}.$$

Define an operator W as follows

$$W : h_k \to k^{-1} f$$
$$W : g \to 0 \quad \text{if} \quad g \perp h_k \quad \text{for} \quad k = 1,2,\ldots .$$

Clearly W is a well defined, bounded linear operator and $\|W\| \leq 2$. Thus

$$\|A^{m_k} W A^{-m_k}\| \geq \|A^{m_k} W A^{-m_k} h_k\| \geq \|A^{m_k}(k^{-1} f + W e_k)\|$$
$$\geq k - 2\|A^{m_k}\|^{-1}\|A^{m_k}\| .$$

Again the right hand side tends to ∞ which contradicts the hypothesis. Thus we have shown that A is power bounded. Since $(A^k T A^{-k})^* = A^{*-k} T^* A^k$, the same proof may be applied to show that $(A^*)^{-1}$ or A^{-1} is power bounded. Since $\|A^k\| \leq M$ for $k = 0, \pm 1, \pm 2,$ it follows from a theorem of Sz. Nagy [4], that A is similar to a unitary operator which completes the proof.

REMARK. By being a bit more fastidious about the epsilontics one can show that if $\sup_n \|A^n K A^{-n}\| \leq M_K < \infty$ for all compact operators $K \in \mathcal{L}(\mathcal{H})$, then again A is similar to a scalar multiple of a unitary operator. Indeed, it suffices to know merely that $\sup_n \|A^n F A^{-n}\| \leq M_F < \infty$ for all rank one operators F. We would like to consider Deddens' question for the world's most popular C^* algebra, the Calkin algebra. (If $\mathcal{K}$ is the ideal of compact operators in $\mathcal{L}(\mathcal{H})$, then the Calkin algebra, which will be denoted by $\mathfrak{A}$ for the rest of the paper, is $\mathcal{L}(\mathcal{H})/\mathcal{K}$).

LEMMA 1. Let $a \in \mathfrak{A}$. Then there exist mutually orthogonal projections $\{p_i\}$ $i = 1,2,\ldots$; such that $\|a^k p_k\| \geq \|a^k\| - k^{-1}$ for $k = 1,2,\ldots$.

<u>Proof</u>. Choose $A \in \mathcal{L}(\mathcal{H})$ such that $\pi(A) = a$. Fix ε, k and a finite dimensional subspace $\mathfrak{M}$.

<u>Claim</u>. There exists a unit vector $f \in \mathfrak{M}^\perp$ such that $\|A^k f\| \geq \|a^k\| - \varepsilon$. For if not set $K = P_{\mathfrak{M}} A^k P_{\mathfrak{M}} + P_{\mathfrak{M}}^\perp A^k P_{\mathfrak{M}} + P_{\mathfrak{M}} A^k P_{\mathfrak{M}}^\perp$. Then K is compact and $\|a^k\| \leq \|A^k - K\| = \|A^k \mid P_{\mathfrak{M}}^\perp \mathcal{H}\| \leq \|a^k\| - \varepsilon$, a contradiction. In virtue of the claim we can choose orthonormal vectors $\{f_{j,k}\}$ $j,k = 1,2,\ldots$ such that

$$\|A^j f_{j,k}\| \geq \|a^j\| - 1/j \quad \text{for} \quad k = 1,2,\ldots .$$

(Having selected $f_{j,k}$ for $j + k \leq n$ set $\mathfrak{M}$ = c.l.m. $\{f_{j,k} : j + k \leq n\}$ and then select $f_{n,1}$ etc.) Set

$$P_j = \text{proj on c.l.m. } \{f_{j,k} : k = 1,2,\ldots\} .$$

For any $K \in \mathcal{K}$,

$$\|(A^j + K)P_j\| \geq \|(A^j + K)P_j f_{j,k}\| \geq \|A^j f_{j,k}\| - \|Kf_{j,k}\|$$
$$\geq \|a^j\| - 1/j - \|Kf_{j,k}\| .$$

Since $\lim_{k\to\infty} \|Kf_{j,k}\| = 0$; it follows that $\|a^j p_j\| \geq \|a^j\| - 1/j$. Clearly $P_i P_j = 0$ for $i \neq j$ whence $p_i p_j = 0$.

THEOREM 2. Let $a \in \mathfrak{A}$ with a invertible. Assume that $\mathfrak{B}_a = \mathfrak{A}$. Then there exist constants α, M such that $\|(\alpha a)^n\| \leq M$ for $n = 0, \pm 1, \pm 2, \ldots$.

Proof. Choose a point $\lambda \in \partial\sigma(a)$. Set $b = \lambda^{-1}a$. Obviously $\mathfrak{B}_a = \mathfrak{A}$ and $1 \in \partial\sigma(b)$. It follows from Theorems 3.1 and 4.1 of [3] that $bq = q$ for some non-zero projection $q \in \mathfrak{A}$. Assume $\|b^n\|$ is not bounded and choose a subsequence n_k such that $\|b^{n_k}\| \geq k^2$. Choose a sequence of projections $\{p_k\}$ in accordance with the previous Lemma so that

$$\|b^{n_k} p_k\| > k^2 - 1/k$$

and

$$p_i p_j = 0 \quad \text{for} \quad i \neq j .$$

Choose an $x \in \mathfrak{A}$ such that $xq = \bigoplus_1^\infty k^{-1} p_k$. (This simply amounts to choosing an $X \in \mathcal{L}(\mathcal{H})$ such that $XQ = \oplus k^{-1} P_k$, which is not difficult.) Then

$$\|b^{n_k} x\, b^{-n_k}\| \geq \|b^{n_k} x\, b^{-n_k} q\| = \|b^{n_k} x\, q\| = \|b^{n_k}[\oplus k^{-1} p_k]\|$$
$$\geq k^{-1}\|b^{n_k} p_k\| \geq (k-1) .$$

Since the right hand side becomes unbounded this contradicts the hypothesis and hence $\|b^n\| \leq M$ for some constant M. To handle the negative powers of b we note that $1 \in \partial\sigma(b^{-1})$ and $(b^n x b^{-n})^* = b^{*-n} x^* b^{*n}$. Thus the same argument shows that $\|b^{*n}\| = \|b^n\|$ is uniformly bounded in n which completes the proof.

COROLLARY. Let $\pi : \mathfrak{A} \to \mathcal{L}(\mathcal{H}_0)$ be a nondegenerate *-representation of the Calkin algebra. Assume that $\mathfrak{B}_a = \mathfrak{A}$ for some $a \in \mathfrak{A}$. Then $\pi(a)$ is similar to a scalar multiple of a unitary operator.

Question. Given $a \in \mathfrak{A}$ where $\|a^k\| \leq M$ for $k = 0, \pm 1, \pm 2, \ldots$ is a similar to a unitary element of $\mathfrak{A}$? It is not obvious (to us, at least) how to generalize Sz. Nagy's Theorem to a C^* algebra, where there is no weak topology readily available.

Note. The first theorem of this paper was also proved by D. A. Herrero using very different techniques.

REFERENCES

[1] J. A. Deddens, Another description of nest algebras (this conference).

[2] J. A. Deddens and T. K. Wong, The commutant of analytic Toeplitz operators, Trans. Amer. Math. Soc., 184 (1973), 261-273.

[3] P. A. Fillmore, J. G. Stampfli and J. P. Williams, On the numerical range, the essential spectrum, and a problem of Halmos, Acta Sci. Math. (Szeged), 33 (1972), 179-192.

[4] B. Sz. Nagy, On uniformly bounded linear transformations in Hilbert space, Acta Sci. Math. (Szeged), 11 (1947), 87-92.

INDIANA UNIVERSITY
BLOOMINGTON, INDIANA

THE FUGLEDE COMMUTATIVITY THEOREM MODULO THE HILBERT-SCHMIDT CLASS AND GENERATING FUNCTIONS FOR MATRIX OPERATORS

Gary Weiss

We have obtained the following results on bounded linear operators acting on a separable, infinite-dimensional, complex Hilbert space. (1) For every normal operator N and every bounded operator X, the Hilbert-Schmidt norms (finite or infinite) of $NX - XN$ and $N^*X - XN^*$ are equal. In particular, if $NX = XN$ modulo C_2, then $N^*X = XN^*$ modulo C_2. (2) If X is a Hilbert-Schmidt operator and N is a normal operator so that $NX - XN$ is a trace class operator, then $\mathrm{Trace}(NX - XN) = 0$. (3) Every normal operator that is similar to a Hilbert-Schmidt perturbation of a diagonal operator D is also unitarily equivalent to a Hilbert-Schmidt perturbation of D. The main technique employs the use of a concept which we call 'generating functions for matrices'.

The second and the third result together with partial results on the first will appear in the Trans. Amer. Math. Soc. under the above title. A sequel is in preparation. It will contain the first result and related ones.

Let H denote a separable, complex Hilbert space and let $L(H)$ denote the class of all bounded linear operators acting on H. Let $K(H)$ denote the class of compact operators in $L(H)$ and let C_p denote the Schatten p-class $(0 < p < \infty)$ with $\|\cdot\|_p$ $(1 \leq p < \infty)$ denoting the associated p-norm. Hence C_2 is the Hilbert-Schmidt class of C_1 is the trace class.

Consider the following statements:

(1) For every normal operator N and $\varepsilon > 0$, there exists a diagonal operator D and a Hilbert-Schmidt operator K_ε with $\|K_\varepsilon\|_2 < \varepsilon$ for which $N \cong D + K_\varepsilon$ ($\cong$ denotes unitary equivalence),

(2) For every normal operator N, there exists a diagonal operator D and a $K \in C_2$ for which $N \cong D + K$,

(3) For every normal operator N and bounded operator X, $\|NX - XN\|_2 = \|N^*X - XN^*\|_2$,

(4) For every normal operator N and bounded operator X, $NX - XN \in C_2$ implies $N^*X - XN^* \in C_2$,

(5) For every normal operator N and bounded operator X, if $NX - XN \in C_2$ and $N^*X - XN^* \in C_2$, then $\|NX - XN\|_2 = \|N^*X - XN^*\|_2$,

(6) If N is normal, $X \in C_2$, and $NX - XN \in C_1$, then Trace $(NX - XN) = 0$.

In [13] Weyl proved that every self-adjoint operator is a compact perturbation of a diagonalizable operator, and that the perturbation may be chosen with an arbitrarily small operator norm. In [8], von Neumann proved that the perturbation could be chosen to be in the Hilbert-Schmidt class and with arbitrarily small Hilbert-Schmidt norm. He proved this in order to obtain results about integral operators. In [1], I. D. Berg generalized Weyl's result to normal operators, and proved that if the spectrum of the normal operator is 'thin enough', then the compact perturbation can also be chosen to be a Hilbert-Schmidt operator with an arbitrarily small Hilbert-Schmidt norm. He asked whether or not the von Neumann result generalizes to all normal operators (that is, statements (1) and (2)). These questions remain open. He conjectured that the full generalization fails and that he believes a barrier preventing a normal operator from having the representation (1) or (2) is that its absolutely continuous part have a spectrum of positive 2-dimensional Lebesque measure. At present, not a single such normal operator is known which can be represented as in (1) or (2).

The 1970's has seen a flurry of deep results on the perturbation theory of operators and the theory of commutators. Besides Berg's paper [1], some of the well-known papers relating perturbation theory to commutators are Berger and Shaw [2], Brown, Douglas and Fillmore [3], Carey and Pincus [4], and Helton and Howe [7].

The connection between (3)-(6) and the Berg problem (2) is clear from the next remarks.

The following implications hold true.

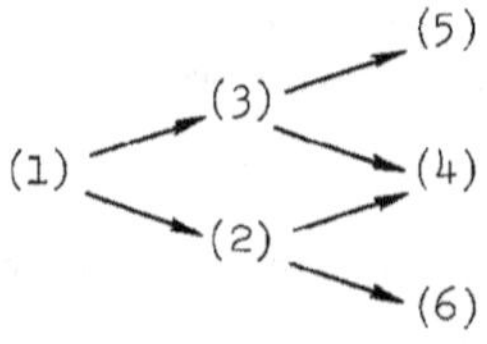

Their proofs are elementary and fairly well-known so we omit them (see [10] or [9, p. 154-162]).

Recently, Donald Hadwin has pointed out to me that (2) → (1). His techniques are different from the mainstream of ideas here, so we omit the details.

The main results of this paper are that statements (3)-(6) are true. Note that statement (4) is the Fuglede Commutativity Theorem modulo the Hilbert-Schmidt class. We shall state the definitions, lemmas, theorems and corollaries and in important cases, we shall summarize the main ideas of the proofs when possible. A complete account of this material will appear in [11], except for statements (3)-(4). They are proved in [12] along with other results.

Definition.

A Laurent operator is an operator of the form M_φ acting on $L^2(T)$, where $\varphi(z) \in L^\infty(T)$ and T denotes the unit circle.

Definition.

If N is a normal operator and $\varphi(z) \in L^\infty(T)$, then M_φ is called a Laurent part of N provided M_φ has no eigenvectors and there exists a diagonal operator D such that $N \cong M_\varphi \oplus D$.

LEMMA 1. Every normal operator is the direct sum of a diagonalizable operator and a Laurent part.

Sketch of proof. Write $N \cong D \oplus N_1$, where D is the diagonalizable operator given by restricting N to the span of its eigenvectors. Then N_1 is normal with no eigenvectors. The spectral theorem realizes N_1 as a multiplication operator relative to a finite non-atomic measure space. The Caratheodory Measure Isomorphism Theorem [6] allows us to replace this measure space by the circle with Lebesgue measure.

This lemma provides us with a crucial canonical form for the commutator $NX - XN$. Letting X be any operator in $L(H)$, relative to $H = M \oplus M^\perp$, where M denotes the domain of D, we obtain

$$N = \begin{pmatrix} D & 0 \\ 0 & M_\varphi \end{pmatrix} \quad \text{and} \quad X = \begin{pmatrix} X_1 & X_2 \\ X_3 & X_4 \end{pmatrix}.$$

A computation then shows that

$$\text{(I)} \qquad NX - XN = \begin{pmatrix} DX_1 - X_1D & DX_2 - X_2M_\varphi \\ M_\varphi X_3 - X_3D & M_\varphi X_4 - X_4M_\varphi \end{pmatrix}$$

and

$$\text{(II)} \qquad N^*X - XN^* = \begin{pmatrix} D^*X_1 - X_1D^* & D^*X_2 - X_2M_\varphi^* \\ M_\varphi^*X_3 - X_3D^* & M_\varphi^*X_4 - X_4M_\varphi^* \end{pmatrix}.$$

Clearly then

$$\text{(I')} \quad \|NX - XN\|_2^2 = \|DX_1 - X_1D\|_2^2 + \|DX_2 - X_2M_\varphi\|_2^2 + \|M_\varphi X_3 - X_3D\|_2^2 + \|M_\varphi X_4 - X_4M_\varphi\|_2^2;$$

and

$$\text{(II')} \quad \|N^*X - XN^*\|_2^2 = \|D^*X_1 - X_1D^*\|_2^2 + \|D^*X_2 - X_2M_\varphi^*\|_2^2 + \|M_\varphi^*X_3 - X_3D^*\|_2^2 + \|M_\varphi^*X_4 - X_4M_\varphi^*\|_2^2.$$

The following theorem relates (I'), (II') and (I) to statements (3)-(6).

THEOREM 2. (a) For every diagonalizable operator D, and $X \in L(H)$, $\|DX - XD\|_2 = \|D^*X - XD^*\|_2$.

(b) For every normal operator N, diagonalizable operator D, and $X \in L(H)$, $\|DX - XN\|_2 = \|D^*X - XN^*\|_2$ and $\|XD - NX\|_2 = \|XD^* - N^*X\|_2$.

(c) To prove any of the statements (1)-(6), it is necessary and sufficient to prove the corresponding statement for the special case when $N = M_\varphi$, where $\varphi(z) \in L^\infty(T)$ and $H = L^2(T)$.

REMARK. A simple computation proves part (a). It may also be found in [9, p. 147] or [10]. To prove part (c), consider separately each of the statements (1)-(6). To obtain (1)-(2), use the Laurent decomposition for a normal operator. To obtain (3)-(5) consider (I') and (II') and to obtain (6) consider (I). The proof of part (b) is not so easy. It requires some techniques developed in [9,

p. 147, Theorem 8e]. We omit the proof.

Before developing the main technique, we are able to obtain a Corollary to Theorem 2 that bears directly on the problem of I. D. Berg (statement (2)). Recall that $\cong$ denotes unitary equivalance and $\sim$ denotes similarity. Berg's problem asks if for every normal operator N, there exists a diagonalizable operator D, and $K \in C_2$ such that $N \cong D + K$. The next corollary shows that Berg's problem is equivalent to the corresponding problem relative to similarity.

COROLLARY 3. If N is a normal operator and D is a diagonalizable operator, then $N \cong D + K$ for some $K \in C_2$ if and only if $N \sim D + K_1$ for some $K_1 \in C_2$.

REMARK. The proof of this corollary follows a similar line as the proof of the well-known fact that similar normal operators are unitarily equivalent. In addition, we need to employ the Weyl-von Neumann Theorem [8] for self-adjoint operators and Theorem 2(b). We omit the proof.

THE MAIN CONSTRUCTION

In this construction, we use the notation that was introduced earlier.

By virtue of Theorem 2(c) we devote our attention to $M_\varphi X - XM_\varphi$, where $\varphi \in L^\infty(T)$ and $X \in L(L^2(T))$. In addition, if $NX - XN$ is a trace class operator and $X \in C_2$, then from the matrix computation I, it is easy to see that (for $N \cong D \oplus M_\varphi$)$M_\varphi X_4 - X_4 M_\varphi$ must be a trace class operator with $X_4 \in C_2$, and $\mathrm{Trace}(NX - XN) = \mathrm{Trace}(M_\varphi X_4 - X_4 M_\varphi)$ (since $\mathrm{Trace}(DX_1 - X_1 D) = 0$).

Note that the matrix for M_φ is a Laurent matrix. Its entries are constant on the diagonals, and those constants are the Fourier coefficients, $\{\varphi_n\}$, of φ.

Let us now introduce generating functions for matrix operators. They are related to Schwartz kerenels in distribution theory.

<u>Definition</u>.

Let $X = (x_{ij}) \in L(L^2(T))$. The <u>generating function</u> for X is defined as the formal Fourier series given by $F(z,w) = \sum_{i,j=-\infty}^{\infty} x_{ij} z^i w^j$.

It is easy to see that the following operation is well-defined, and that it possesses certain obvious distributive and associative properties.

Definition.

Let $\varphi, \psi \in L^\infty(T)$ where $\varphi(z) = \sum_n \varphi_n z^n$ and $\psi(z) = \sum_n \psi_n z^n$, and let $X \in L(L^2(T))$ so that $F(z,w) = \sum_{i,j} x_{ij} z^i w^j$ is the generating function for X. Define the binary operation $*$ as follows

$$[\varphi(z) + \psi(w)] * F(z,w) = \sum_{i,j} \left(\sum_n (\varphi_n x_{i-n,j} + \psi_n x_{i,j-n}) \right) z^i w^j .$$

It is helpful to recognize that $*$ simply denotes the formal product of these power series and that this same symbol is used to denote formal products in some computer languages.

Also the reader should take care not to confuse this symbol with the symbol for operator adjoints.

Let us now compute the generating function for $M_\varphi X - XM_\varphi$.

$$(M_\varphi X)_{i,j} = \left((\varphi_{j-i})(x_{ij}) \right)_{i,j} = \sum_k \varphi_{k-i} x_{kj} = \sum_n \varphi_n x_{i+n,j}$$

and

$$(XM_\varphi)_{i,j} = \sum_k x_{ik} \varphi_{j-k} = \sum_n \varphi_n x_{i,j-n} .$$

Also, $M_\varphi^* = M_{\varphi^*}$, where $\varphi^*(z) = \sum_n \bar{\varphi}_{-n} z^n$, and $(M_\varphi^*)_{i,j} = \bar{\varphi}_{i-j}$. This gives us the following information about $M_\varphi^* X - XM_\varphi^*$.

$$(M_\varphi^* X)_{i,j} = \left((\bar{\varphi}_{i-j})(x_{ij}) \right)_{i,j} = \sum_k \bar{\varphi}_{i-k} x_{kj} = \sum_n \bar{\varphi}_n x_{i-n,j}$$

and

$$(XM_\varphi^*)_{i,j} = \sum_k x_{ik} \bar{\varphi}_{k-j} = \sum_n \bar{\varphi}_n x_{i,j+n} .$$

So $(M_\varphi X - XM_\varphi)_{i,j} = \sum_n \varphi_n (x_{i+n,j} - x_{i,j-n})$ and $(M_\varphi^* X - XM_\varphi^*)_{i,j} = \sum_n \bar{\varphi}_n (x_{i-n,j} - x_{i,j+n})$.

Now regard $F(z,w) = \sum x_{ij} z^i w^j$ as a distribution on $C^\infty(T^2)$. Then a computation shows

$$x_{i+n,j} = \langle \bar{z}^n F, z^k w^j \rangle, \quad x_{i,j-n} = \langle w^n F, z^i w^j \rangle$$

$$x_{i-n,j} = \langle z^n F, z^i w^j \rangle, \quad x_{i,j+n} = \langle \bar{w}^n F, z^i w^j \rangle .$$

An additional computation shows

$$(M_\varphi X - XM_\varphi)_{i,j} = \sum_n \varphi_n \langle (\overline{z}^n - w^n)F, z^i w^j \rangle = \langle (\varphi(\overline{z}) - \varphi(w)) * F, z^i w^j \rangle,$$

and

$$(M_\varphi^* X - XM_\varphi^*)_{i,j} = \sum_n \overline{\varphi}_n \langle (z^n - \overline{w}^n)F, z^i w^j \rangle = \langle \overline{(\varphi(\overline{z}) - \varphi(w))} * F, z^i w^j \rangle .$$

This says that the <u>generating function for</u> $M_\varphi X - XM_\varphi$ <u>is</u> $(\varphi(\overline{z}) - \varphi(w)) * F(z,w)$ <u>and the generating function for</u> $M_\varphi^* X - XM_\varphi^*$ <u>is</u> $\overline{(\varphi(\overline{z}) - \varphi(w))} * F(z,w)$. (Note, the equalities above are best proven by computing the last expressions first, in terms of φ_n and x_{ij}.)

This completes the construction of the generating function for the commutators. We can now prove statement (5).

FUGLEDE'S THEOREM MODULO C_2

THEOREM 4. If N is normal, $X \in L(H)$, and $NX - XN \in C_2$ and $N^*X - XN^* \in C_2$, then $\|NX - XN\|_2 = \|N^*X - XN^*\|_2$.

<u>Proof</u>. By Theorem 2(c) it suffices to prove Theorem 4 when $N = M_\varphi$ acting on $L^2(T)$, with $\varphi \in L^\infty(T)$, and such that M_φ has no eigenvalues. By the main construction, the assumption on M_φ and X is equivalent to the assumptions that $(\varphi(\overline{z}) - \varphi(w) * F(z,w) \in L^2(T^2)$ and $\overline{(\varphi(\overline{z}) - \varphi(w))} * F(z,w) \in L^2(T^2)$. Of course, here we are treating those formal Fourier series in z, w which have square summable coefficients as functions in $L^2(T^2)$. Therefore the entries of $M_\varphi X - XM_\varphi$ are the coefficients of $(\varphi(\overline{z}) - \varphi(w)) * F(z,w)$, and by Bessel's equality, satisfy

$$\|M_\varphi X - XM_\varphi\|_2^2 = \sum_{i,j} |(M_\varphi X - XM_\varphi)_{i,j}|^2 = \iint_{T^2} |(\varphi(\overline{z}) - \varphi(w)) * F(z,w)|^2 .$$

Similarly,

$$\|M_\varphi^* X - XM_\varphi^*\|_2^2 = \iint_{T^2} |\overline{(\varphi(\overline{z}) - (w))} * F(z,w)|^2 .$$

Now the assumption that M_φ has no eigenvectors is needed. It guarantees that $\varphi(\overline{z}) \neq \varphi(w)$ almost everywhere with respect to 2-dimensional Lebesgue measure on T^2.

Using this function we obtain

$$(*) = \iint_{T^2} |(\varphi(\bar{z}) - \varphi(w)) * F(z,w)|^2 = \iint_{T^2} \left| \frac{\overline{\varphi(\bar{z}) - \varphi(w)}}{\varphi(\bar{z}) - \varphi(w)} ((\varphi(\bar{z}) - \varphi(w)) * F(z,w)) \right|^2$$

$$= \iint_{T^2} \left| \frac{1}{\varphi(\bar{z}) - \varphi(w)} \overline{(\varphi(\bar{z}) - \varphi(w))} ((\varphi(\bar{z}) - \varphi(w)) * F(z,w)) \right|^2 .$$

In addition, for every normal operator N, the derivations δ_N and δ_{N^*} commute (the proof is simple algebra). Hence,

$$M_\varphi^*(M_\varphi X - XM_\varphi) - (M_\varphi X - XM_\varphi) M_\varphi^* = M_\varphi(M_\varphi^* X - XM_\varphi^*) - (M_\varphi^* X - XM_\varphi^*) M_\varphi .$$

The generating function for the left hand side of this equality is given by $(\varphi(\bar{z}) - \varphi(w)) * ((\varphi(\bar{z}) - \varphi(w)) * F(z,w))$, which is the same formal Fourier series as $\overline{(\varphi(\bar{z}) - \varphi(w))} ((\varphi(\bar{z}) - \varphi(w)) * F(z,w))$ because of the assumption that $(\varphi(\bar{z}) - \varphi(w)) * F(z,w)$ is a function in $L^2(T^2)$. Similarly, the generating function for the right hand side of the equality is given by $(\varphi(\bar{z}) - \varphi(w)) * \overline{((\varphi(\bar{z}) - \varphi(w)) * F(z,w))} = (\varphi(\bar{z}) - \varphi(w)) \overline{((\varphi(\bar{z}) - \varphi(w))} * F(z,w))$. This last equality follows from the assumption that $\overline{(\varphi(\bar{z}) - \varphi(w))} * F(z,w)$ is also a function in $L^2(T^2)$. Hence $((\varphi(\bar{z}) - \varphi(w)) \overline{((\varphi(\bar{z})} - \varphi(w)) * F(z,w))$ is a power series identical to $(\varphi(\bar{z}) - \varphi(w)) \overline{((\varphi(\bar{z}} - (\varphi(w)) * F(z,w))$. Thus

$$(*) = \int_{T^2} \left| \frac{1}{\varphi(\bar{z}) - \varphi(w)} (\varphi(\bar{z}) - \varphi(w)) \overline{((\varphi(\bar{z}) - \varphi(w)) * F(z,w))} \right|^2$$

$$= \int_{T^2} |(\varphi(\bar{z}) - \varphi(w)) * F(z,w)|^2 .$$

Whence $\|M_\varphi X - XM_\varphi\|_2^2 = \|M_\varphi^* X - XM_\varphi^*\|_2^2$.

Using these ideas and deeper analysis, we obtain the following results.

THEOREM 5. If N is normal and $X \in L(H)$, then the Hilbert-Schmidt norms (finite or infinite) of $NX - XN$ and $N^*X - XN^*$ are equal. In particular,

$NX - XN \in C_2$ implies $N^*X - XN^* \in C_2$.

THE TRACE OF $NX - XN$

Here we assume N is a normal operator and X is a Hilbert-Schmidt operator for which $NX - XN$ is a trace class operator. The techniques relating to this question are too complicated to describe here. We simply list the results. The deeper results require Lusin's Theorem, Egoroff's Theorem and convergence in measure.

THEOREM 6. If $\varphi(x) \in L^\infty[0,1]$ then $$\iint_{[0,1]\times[0,1]} \frac{1}{|\varphi(x)-\varphi(y)|^2}\, dxdy = \infty .$$

COROLLARY 7. If E is any measurable subset of $[0,1]$ of positive measure, and $|\varphi|$ is essentially bounded on E, then $$\iint_{E\times E} \frac{1}{|\varphi(x)-\varphi(y)|^2}\, dxdy = \infty .$$

THEOREM 8. If N is a normal operator, $X \in C_2$ and $NX - XN \in C_1$, then $\operatorname{trace}(NX - XN) = 0$.

THEOREM 9. Let $\varphi \in H^\infty(T)$ and let T_φ denote the analytic Toeplitz operator with symbol φ. Then the following two statements are equivalent.

(a) $(T_\varphi)' \cap C_2 \neq \{0\}$.

(b) There exists $F(z,w) = \sum_{i,j=1}^{\infty} x_{ij} z^i w^j$ with $\sum |x_{ij}|^2 < \infty$ such that $(\varphi(z) - \varphi(w))F(\bar{z},w) \in H^2(T^2)$.

ACKNOWLEDGMENTS

I should like to acknowledge the following mathematicians for their helpful suggestions and encouragements: Allen Shields, Jeffrey Rauch, Hugh Montgomery and Joel Anderson.

REFERENCES

[1] I. D. Berg, An extension of the Weyl-von Neumann Theorem to normal operators, Trans. Amer. Math. Soc. 160 (1971), 365-371.

[2] C. A. Berger and B. I. Shaw, Self-commutators of multicyclic hyponormal operators are always trace class, Bull. Amer. Math. Soc. 79 (1973), 1193-1199.

[3] L. G. Brown, R. G. Douglas, and P. A. Fillmore, Unitary equivalence modulo the compact operators and extensions of C*-algebras, Springer-Verlag, 345 (1973), 58-128.

[4] R. W. Carey and J. D. Pincus, Perturbation by trace class operators, Bull. Amer. Math. Soc. 80 (1974), 758-759.

[5] I. C. Gohberg and M. G. Krein, Introduction to the Theory of Linear Nonself-adjoint Operators. Translated from the Monographs, Vol. 18, American Mathematical Society, Providence, Rhode Island, 1969.

[6] P. R. Halmos, Measure Theory, D. van Nostrand Co. (1950), 171-174.

[7] J. W. Helton and R. E. Howe, Integral operators: commutators, traces, index and homology, Springer-Verlag, 345 (1973), 141-209.

[8] J. von Neumann, Charakterisierung des Spektrums eines Integraloperators, Actualites Sci. Indust., no. 229, Hermann, Paris, 1935.

[9] Gary Weiss, Commutators and operator ideals, dissertation (1975), University of Michigan.

[10] Gary Weiss, Fuglede's Commutativity Theorem modulo operator ideals, submitted.

[11] Gary Weiss, The Fuglede Commutativity Theorem Modulo the Hilbert-Schmidt Class and Generating Functions for Matrix Operators I, Trans. Amer. Math. Soc., to appear.

[12] Gary Weiss, The Fuglede Commutativity Theorem Modulo the Hilbert-Schmidt Class and Generating Functions for Matrix Operators II, in preparation.

[13] H. Weyl, "Über beschränkte quadratischen Formen deren Differenz vollstetig ist," Rend. Circ. Mat. Pelermo, 27 (1909), 373-392.

UNIVERSITY OF CINCINATTI

Vol. 521: G. Cherlin, Model Theoretic Algebra – Selected Topics. IV, 234 pages. 1976.

Vol. 522: C. O. Bloom and N. D. Kazarinoff, Short Wave Radiation Problems in Inhomogeneous Media: Asymptotic Solutions. V. 104 pages. 1976.

Vol. 523: S. A. Albeverio and R. J. Høegh-Krohn, Mathematical Theory of Feynman Path Integrals. IV, 139 pages. 1976.

Vol. 524: Séminaire Pierre Lelong (Analyse) Année 1974/75. Edité par P. Lelong. V, 222 pages. 1976.

Vol. 525: Structural Stability, the Theory of Catastrophes, and Applications in the Sciences. Proceedings 1975. Edited by P. Hilton. VI, 408 pages. 1976.

Vol. 526: Probability in Banach Spaces. Proceedings 1975. Edited by A. Beck. VI, 290 pages. 1976.

Vol. 527: M. Denker, Ch. Grillenberger, and K. Sigmund, Ergodic Theory on Compact Spaces. IV, 360 pages. 1976.

Vol. 528: J. E. Humphreys, Ordinary and Modular Representations of Chevalley Groups. III, 127 pages. 1976.

Vol. 529: J. Grandell, Doubly Stochastic Poisson Processes. X, 234 pages. 1976.

Vol. 530: S. S. Gelbart, Weil's Representation and the Spectrum of the Metaplectic Group. VII, 140 pages. 1976.

Vol. 531: Y.-C. Wong, The Topology of Uniform Convergence on Order-Bounded Sets. VI, 163 pages. 1976.

Vol. 532: Théorie Ergodique. Proceedings 1973/1974. Edité par J.-P. Conze and M. S. Keane. VIII, 227 pages. 1976.

Vol. 533: F. R. Cohen, T. J. Lada, and J. P. May, The Homology of Iterated Loop Spaces. IX, 490 pages. 1976.

Vol. 534: C. Preston, Random Fields. V, 200 pages. 1976.

Vol. 535: Singularités d'Applications Differentiables. Plans-sur-Bex. 1975. Edité par O. Burlet et F. Ronga. V, 253 pages. 1976.

Vol. 536: W. M. Schmidt, Equations over Finite Fields. An Elementary Approach. IX, 267 pages. 1976.

Vol. 537: Set Theory and Hierarchy Theory. Bierutowice, Poland 1975. A Memorial Tribute to Andrzej Mostowski. Edited by W. Marek, M. Srebrny and A. Zarach. XIII, 345 pages. 1976.

Vol. 538: G. Fischer, Complex Analytic Geometry. VII, 201 pages. 1976.

Vol. 539: A. Badrikian, J. F. C. Kingman et J. Kuelbs, Ecole d'Eté de Probabilités de Saint Flour V-1975. Edité par P.-L. Hennequin. IX, 314 pages. 1976.

Vol. 540: Categorical Topology, Proceedings 1975. Edited by E. Binz and H. Herrlich. XV, 719 pages. 1976.

Vol. 541: Measure Theory, Oberwolfach 1975. Proceedings. Edited by A. Bellow and D. Kölzow. XIV, 430 pages. 1976.

Vol. 542: D. A. Edwards and H. M. Hastings, Čech and Steenrod Homotopy Theories with Applications to Geometric Topology. VII, 296 pages. 1976.

Vol. 543: Nonlinear Operators and the Calculus of Variations, Bruxelles 1975. Edited by J. P. Gossez, E. J. Lami Dozo, J. Mawhin, and L. Waelbroeck, VII, 237 pages. 1976.

Vol. 544: Robert P. Langlands, On the Functional Equations Satisfied by Eisenstein Series. VII, 337 pages. 1976.

Vol. 545: Noncommutative Ring Theory. Kent State 1975. Edited by J. H. Cozzens and F. L. Sandomierski. V, 212 pages. 1976.

Vol. 546: K. Mahler, Lectures on Transcendental Numbers. Edited and Completed by B. Diviš and W. J. Le Veque. XXI, 254 pages. 1976.

Vol. 547: A. Mukherjea and N. A. Tserpes, Measures on Topological Semigroups: Convolution Products and Random Walks. V, 197 pages. 1976.

Vol. 548: D. A. Hejhal, The Selberg Trace Formula for PSL (2,ℝ). Volume I. VI, 516 pages. 1976.

Vol. 549: Brauer Groups, Evanston 1975. Proceedings. Edited by D. Zelinsky. V, 187 pages. 1976.

Vol. 550: Proceedings of the Third Japan – USSR Symposium on Probability Theory. Edited by G. Maruyama and J. V. Prokhorov. VI, 722 pages. 1976.

Vol. 551: Algebraic K-Theory, Evanston 1976. Proceedings. Edited by M. R. Stein. XI, 409 pages. 1976.

Vol. 552: C. G. Gibson, K. Wirthmüller, A. A. du Plessis and E. J. N. Looijenga. Topological Stability of Smooth Mappings. V, 155 pages. 1976.

Vol. 553: M. Petrich, Categories of Algebraic Systems. Vector and Projective Spaces, Semigroups, Rings and Lattices. VIII, 217 pages. 1976.

Vol. 554: J. D. H. Smith, Mal'cev Varieties. VIII, 158 pages. 1976.

Vol. 555: M. Ishida, The Genus Fields of Algebraic Number Fields. VII, 116 pages. 1976.

Vol. 556: Approximation Theory. Bonn 1976. Proceedings. Edited by R. Schaback and K. Scherer. VII, 466 pages. 1976.

Vol. 557: W. Iberkleid and T. Petrie, Smooth S^1 Manifolds. III, 163 pages. 1976.

Vol. 558: B. Weisfeiler, On Construction and Identification of Graphs. XIV, 237 pages. 1976.

Vol. 559: J.-P. Caubet, Le Mouvement Brownien Relativiste. IX, 212 pages. 1976.

Vol. 560: Combinatorial Mathematics, IV, Proceedings 1975. Edited by L. R. A. Casse and W. D. Wallis. VII, 249 pages. 1976.

Vol. 561: Function Theoretic Methods for Partial Differential Equations. Darmstadt 1976. Proceedings. Edited by V. E. Meister, N. Weck and W. L. Wendland. XVIII, 520 pages. 1976.

Vol. 562: R. W. Goodman, Nilpotent Lie Groups: Structure and Applications to Analysis. X, 210 pages. 1976.

Vol. 563: Séminaire de Théorie du Potentiel. Paris, No. 2. Proceedings 1975–1976. Edited by F. Hirsch and G. Mokobodzki. VI, 292 pages. 1976.

Vol. 564: Ordinary and Partial Differential Equations, Dundee 1976. Proceedings. Edited by W. N. Everitt and B. D. Sleeman. XVIII, 551 pages. 1976.

Vol. 565: Turbulence and Navier Stokes Equations. Proceedings 1975. Edited by R. Temam. IX, 194 pages. 1976.

Vol. 566: Empirical Distributions and Processes. Oberwolfach 1976. Proceedings. Edited by P. Gaenssler and P. Révész. VII, 146 pages. 1976.

Vol. 567: Séminaire Bourbaki vol. 1975/76. Exposés 471–488. IV, 303 pages. 1977.

Vol. 568: R. E. Gaines and J. L. Mawhin, Coincidence Degree, and Nonlinear Differential Equations. V, 262 pages. 1977.

Vol. 569: Cohomologie Etale SGA 4½. Séminaire de Géométrie Algébrique du Bois-Marie. Edité par P. Deligne. V, 312 pages. 1977.

Vol. 570: Differential Geometrical Methods in Mathematical Physics, Bonn 1975. Proceedings. Edited by K. Bleuler and A. Reetz. VIII, 576 pages. 1977.

Vol. 571: Constructive Theory of Functions of Several Variables, Oberwolfach 1976. Proceedings. Edited by W. Schempp and K. Zeller. VI. 290 pages. 1977

Vol. 572: Sparse Matrix Techniques, Copenhagen 1976. Edited by V. A. Barker. V, 184 pages. 1977.

Vol. 573: Group Theory, Canberra 1975. Proceedings. Edited by R. A. Bryce, J. Cossey and M. F. Newman. VII, 146 pages. 1977.

Vol. 574: J. Moldestad, Computations in Higher Types. IV, 203 pages. 1977.

Vol. 575: K-Theory and Operator Algebras, Athens, Georgia 1975. Edited by B. B. Morrel and I. M. Singer. VI, 191 pages. 1977.

Vol. 576: V. S. Varadarajan, Harmonic Analysis on Real Reductive Groups. VI, 521 pages. 1977.

Vol. 577: J. P. May, E_∞ Ring Spaces and E_∞ Ring Spectra. IV, 268 pages. 1977.

Vol. 578: Séminaire Pierre Lelong (Analyse) Année 1975/76. Edité par P. Lelong. VI, 327 pages. 1977.

Vol. 579: Combinatoire et Représentation du Groupe Symétrique, Strasbourg 1976. Proceedings 1976. Edité par D. Foata. IV, 339 pages. 1977.

Vol. 580: C. Castaing and M. Valadier, Convex Analysis and Measurable Multifunctions. VIII, 278 pages. 1977.

Vol. 581: Séminaire de Probabilités XI, Université de Strasbourg. Proceedings 1975/1976. Edité par C. Dellacherie, P. A. Meyer et M. Weil. VI, 574 pages. 1977.

Vol. 582: J. M. G. Fell, Induced Representations and Banach *-Algebraic Bundles. IV, 349 pages. 1977.

Vol. 583: W. Hirsch, C. C. Pugh and M. Shub, Invariant Manifolds. IV, 149 pages. 1977.

Vol. 584: C. Brezinski, Accélération de la Convergence en Analyse Numérique. IV, 313 pages. 1977.

Vol. 585: T. A. Springer, Invariant Theory. VI, 112 pages. 1977.

Vol. 586: Séminaire d'Algèbre Paul Dubreil, Paris 1975–1976 (29ème Année). Edited by M. P. Malliavin. VI, 188 pages. 1977.

Vol. 587: Non-Commutative Harmonic Analysis. Proceedings 1976. Edited by J. Carmona and M. Vergne. IV, 240 pages. 1977.

Vol. 588: P. Molino, Théorie des G-Structures: Le Problème d'Equivalence. VI, 163 pages. 1977.

Vol. 589: Cohomologie l-adique et Fonctions L. Séminaire de Géométrie Algébrique du Bois-Marie 1965–66, SGA 5. Edité par L. Illusie. XII, 484 pages. 1977.

Vol. 590: H. Matsumoto, Analyse Harmonique dans les Systèmes de Tits Bornologiques de Type Affine. IV, 219 pages. 1977.

Vol. 591: G. A. Anderson, Surgery with Coefficients. VIII, 157 pages. 1977.

Vol. 592: D. Voigt, Induzierte Darstellungen in der Theorie der endlichen, algebraischen Gruppen. V, 413 Seiten. 1977.

Vol. 593: K. Barbey and H. König, Abstract Analytic Function Theory and Hardy Algebras. VIII, 260 pages. 1977.

Vol. 594: Singular Perturbations and Boundary Layer Theory, Lyon 1976. Edited by C. M. Brauner, B. Gay, and J. Mathieu. VIII, 539 pages. 1977.

Vol. 595: W. Hazod, Stetige Faltungshalbgruppen von Wahrscheinlichkeitsmaßen und erzeugende Distributionen. XIII, 157 Seiten. 1977.

Vol. 596: K. Deimling, Ordinary Differential Equations in Banach Spaces. VI, 137 pages. 1977.

Vol. 597: Geometry and Topology, Rio de Janeiro, July 1976. Proceedings. Edited by J. Palis and M. do Carmo. VI, 866 pages. 1977.

Vol. 598: J. Hoffmann-Jørgensen, T. M. Liggett et J. Neveu, Ecole d'Eté de Probabilités de Saint-Flour VI – 1976. Edité par P.-L. Hennequin. XII, 447 pages. 1977.

Vol. 599: Complex Analysis, Kentucky 1976. Proceedings. Edited by J. D. Buckholtz and T. J. Suffridge. X, 159 pages. 1977.

Vol. 600: W. Stoll, Value Distribution on Parabolic Spaces. VIII, 216 pages. 1977.

Vol. 601: Modular Functions of one Variable V, Bonn 1976. Proceedings. Edited by J.-P. Serre and D. B. Zagier. VI, 294 pages. 1977.

Vol. 602: J. P. Brezin, Harmonic Analysis on Compact Solvmanifolds. VIII, 179 pages. 1977.

Vol. 603: B. Moishezon, Complex Surfaces and Connected Sums of Complex Projective Planes. IV, 234 pages. 1977.

Vol. 604: Banach Spaces of Analytic Functions, Kent, Ohio 1976. Proceedings. Edited by J. Baker, C. Cleaver and Joseph Diestel. VI, 141 pages. 1977.

Vol. 605: Sario et al., Classification Theory of Riemannian Manifolds. XX, 498 pages. 1977.

Vol. 606: Mathematical Aspects of Finite Element Methods. Proceedings 1975. Edited by I. Galligani and E. Magenes. VI, 362 pages. 1977.

Vol. 607: M. Métivier, Reelle und Vektorwertige Quasimartingale und die Theorie der Stochastischen Integration. X, 310 Seiten. 1977.

Vol. 608: Bigard et al., Groupes et Anneaux Réticulés. XIV, 334 pages. 1977.

Vol. 609: General Topology and Its Relations to Modern Analysis and Algebra IV. Proceedings 1976. Edited by J. Novák. XVIII, 225 pages. 1977.

Vol. 610: G. Jensen, Higher Order Contact of Submanifolds of Homogeneous Spaces. XII, 154 pages. 1977.

Vol. 611: M. Makkai and G. E. Reyes, First Order Categorical Logic. VIII, 301 pages. 1977.

Vol. 612: E. M. Kleinberg, Infinitary Combinatorics and the Axiom of Determinateness. VIII, 150 pages. 1977.

Vol. 613: E. Behrends et al., L^p-Structure in Real Banach Spaces. X, 108 pages. 1977.

Vol. 614: H. Yanagihara, Theory of Hopf Algebras Attached to Group Schemes. VIII, 308 pages. 1977.

Vol. 615: Turbulence Seminar, Proceedings 1976/77. Edited by P. Bernard and T. Ratiu. VI, 155 pages. 1977.

Vol. 616: Abelian Group Theory, 2nd New Mexico State University Conference, 1976. Proceedings. Edited by D. Arnold, R. Hunter and E. Walker. X, 423 pages. 1977.

Vol. 617: K. J. Devlin, The Axiom of Constructibility: A Guide for the Mathematician. VIII, 96 pages. 1977.

Vol. 618: I. I. Hirschman, Jr. and D. E. Hughes, Extreme Eigen Values of Toeplitz Operators. VI, 145 pages. 1977.

Vol. 619: Set Theory and Hierarchy Theory V, Bierutowice 1976. Edited by A. Lachlan, M. Srebrny, and A. Zarach. VIII, 358 pages. 1977.

Vol. 620: H. Popp, Moduli Theory and Classification Theory of Algebraic Varieties. VIII, 189 pages. 1977.

Vol. 621: Kauffman et al., The Deficiency Index Problem. VI, 112 pages. 1977.

Vol. 622: Combinatorial Mathematics V, Melbourne 1976. Proceedings. Edited by C. Little. VIII, 213 pages. 1977.

Vol. 623: I. Erdelyi and R. Lange, Spectral Decompositions on Banach Spaces. VIII, 122 pages. 1977.

Vol. 624: Y. Guivarc'h et al., Marches Aléatoires sur les Groupes de Lie. VIII, 292 pages. 1977.

Vol. 625: J. P. Alexander et al., Odd Order Group Actions and Witt Classification of Innerproducts. IV, 202 pages. 1977.

Vol. 626: Number Theory Day, New York 1976. Proceedings. Edited by M. B. Nathanson. VI, 241 pages. 1977.

Vol. 627: Modular Functions of One Variable VI, Bonn 1976. Proceedings. Edited by J.-P. Serre and D. B. Zagier. VI, 339 pages. 1977.

Vol. 628: H. J. Baues, Obstruction Theory on the Homotopy Classification of Maps. XII, 387 pages. 1977.

Vol. 629: W. A. Coppel, Dichotomies in Stability Theory. VI, 98 pages. 1978.

Vol. 630: Numerical Analysis, Proceedings, Biennial Conference, Dundee 1977. Edited by G. A. Watson. XII, 199 pages. 1978.

Vol. 631: Numerical Treatment of Differential Equations. Proceedings 1976. Edited by R. Bulirsch, R. D. Grigorieff, and J. Schröder. X, 219 pages. 1978.

Vol. 632: J.-F. Boutot, Schéma de Picard Local. X, 165 pages. 1978.

Vol. 633: N. R. Coleff and M. E. Herrera, Les Courants Résiduels Associés à une Forme Méromorphe. X, 211 pages. 1978.

Vol. 634: H. Kurke et al., Die Approximationseigenschaft lokaler Ringe. IV, 204 Seiten. 1978.

Vol. 635: T. Y. Lam, Serre's Conjecture. XVI, 227 pages. 1978.

Vol. 636: Journées de Statistique des Processus Stochastiques, Grenoble 1977, Proceedings. Edité par Didier Dacunha-Castelle et Bernard Van Cutsem. VII, 202 pages. 1978.

Vol. 637: W. B. Jurkat, Meromorphe Differentialgleichungen. VII, 194 Seiten. 1978.

Vol. 638: P. Shanahan, The Atiyah-Singer Index Theorem, An Introduction. V, 224 pages. 1978.

Vol. 639: N. Adasch et al., Topological Vector Spaces. V, 125 pages. 1978.